T0328941

Renewing innovation systems in agriculture and food

About CTA

The Technical Centre for Agricultural and Rural Cooperation (CTA) is a joint international institution of the African, Caribbean and Pacific (ACP) Group of States and the European Union (EU). Its mission is to advance food and nutritional security, increase prosperity and encourage sound natural resource management in ACP countries. It provides access to information and knowledge, facilitates policy dialogue and strengthens the capacity of agricultural and rural development institutions and communities.

CTA operates under the framework of the Cotonou Agreement and is funded by the EU.

For more information on CTA, visit www.cta.int

Renewing innovation systems in agriculture and food

How to go towards more sustainability?

edited by:
E. Coudel
H. Devautour
C.T. Soulard
G. Faure
B. Hubert

This book was also published in french by Quae under the title: 'Apprendre à innover dans un monde incertain: concevoir les futures de l'agriculture et de l'agroalimentaire'. Chapters 1, 5, 6, 8, 11, 12 were translated from French by Kim Agrawal.

ISBN 978-90-8686-214-6
e-ISBN: 978-90-8686-768-4
DOI: 10.3920/978-90-8686-768-4

First published, 2013

Wageningen Academic Publishers
P.O. Box 220
6700 AE Wageningen
The Netherlands
www.WageningenAcademic.com
copyright@WageningenAcademic.com

Acknowledgement

This book is the culmination of a long process of coordination, debate and creativity. We wish to thank all the people who gave life to the ISDA 2010 symposium and capitalized on the many discussions that took place.

Thanks to the "Innovation" Research Unit from Montpellier for launching the idea of bringing together different communities working on innovation in agriculture and food and seeing this project through to its completion.

Thanks to the members of the Scientific Committee for bringing their vision, guidance and spirit of synthesis: B. Hubert (Agropolis International, France), F. Aggeri (Ensmp, France), C. Almekinders (WUR, Nederlands), J. Berdegue (Rimisp, Chile), R. Buruchara (Ciat, Colombia), P. Caron (Cirad, France), B. Chevassus-au Louis (Ministry of Agriculture, France), O. Coomes (University of McGill, Canada), S. Dubuisson Quellier (Cnrs, France), M. Fonte (University of Napoli, Italy), A. Hall (Link limited. & CRT-RIU, UK), S. Hisano (University of Kyoto, Japan), K. Hussein (Ocde), P-B. Joly (Inra, France), J. Kirsten (University of Pretoria, South Africa), J-M. Meynard (Inra, France), D. Requier Desjardins (IEP Toulouse, France), P. A. Seck (Africa Rice, Bénin), R. Teulier (Cnrs, France)

Thanks to the Local Scientific Committee for bringing their enthusiasm throughout the process and for aggregating their networks to make this event a success: H. Devautour, E. Coudel, M. Antona, J-M. Barbier, D. Bazile, E. Biénabe, E. Chia, D. Desclaux, G. Faure, P. Gasselin, P. Maizi, M. Piraux, C. Poncet, B. Prevost, S. Ridaura, C. Soulard, L. Temple, L. Temri, J-M. Touzard, B. Triomphe, E. Valette.

Thanks to all those on the back stage who, with their best effectiveness, coordinated all the logistics, communication, travelling: M-F. Chazalette, B. Gillet, N. Kelemen, E. Grégoire, A. Rossard, C. Rollin, Denis Delebecque, T. Erwin, P. Radigon, C. Rawski, P. Lajous-Causse. A special thanks to K. Agrawal and A. Cockle for translating the chapters of this book.

Thanks to those who agreed to bring their energy to facilitate the sessions: C. Albaladejo, M. Barbier, P. Bonnal, M. Bonin, F. Bousquet, C. Bryant, E. Cheyns, N. Cialdella, Y. Chiffoleau, E. Doidy, M. Dosso, S. Dury, C. Ferraton, S. Fournier, C. Gary, F. Goulet, C. Harris, H. Hocde, H. Ilbert, F. Jarrige, G. Kamau, P-Y. Le Gal, V. Mathieu, A. Quinlan, R. Rajalahti, H. Rey-Valette, S. Ridaura, G. Ruivenkamp, D. Sautier, P. Tittonell, E. Torquebiau, A. Torre, S. de Tourdonnet, J. van den Berg, F. Wallet, A. Waters-Bayer.

Thanks to the various donors for their financial support which enabled so many participants from the South to come: CTA (Centre Technique Agricole), IFAD (International Fund for Agricultural Development), Agropolis Fondation, the french Ministry of culture and communication (General Delegation of the french language and languages of France),

Languedoc-Roussillon Region, Agglomeration of Montpellier. Thanks to the sponsorship of the Ministry of Agriculture and Fisheries, Ministry of Foreign and European Affairs, and the Ministry of Ecology, Energy, Sustainable Development and the Sea. Thanks to the World Bank and OECD (Organization of Economic Cooperation and Development) for their participation.

Finally, thanks to the more than 500 participants for having brought their enthusiasm, their ideas, and their willingness to share into ISDA 2010.

Table of contents

Acronyms

AARINENA	Association of Agricultural Research Institutions in the Near East and North Africa
AFD	French Development Agency
AIS	Agricultural Innovation System
AKIS	Agricultural Knowledge and Information Systems
ANT	Actor-Network Theory
ASARECA	Association of Strengthening Agricultural Research in Eastern and Central Africa
BRAC	Bangladesh Rural Advancement Committee
CBD	Convention on Biological Diversity
CGDA	General Council for Agricultural Development, Morocco.
CIRAD	Centre for International Cooperation in Agronomic Research for Development
CMNR	Collaborative Management of Natural Resources
CPERA	Commissions for Planning and Evaluating Agricultural Research
DFID	Department for International Development, UK
ENA	National Administration School of Rabat, Morocco
ENFI	National School of Forestry Engineers, Morocco
FAO	Food and Agriculture Organization of the United Nations
FAP	Fodder Adoption Project
FARA	Forum for Agricultural Research in Africa
FIPS	Farm Input Promotion Services
FRP	Federated Research Projects
GAPs	Good Agricultural Practices
GCARD	Global Conference on Agricultural Research for Development
GCRAI	Consultative Group on International Agricultural Research (*French acronym*)
GDP	Gross Domestic Product
GEF	Global Environment Facility
GIAHs	Globally Important Agricultural Heritage Systems
GMO	Genetically Modified Organisms
GREMI	Group for European Research on Innovative Environments
HDI	Human Development Indicator
IAV	Hassan II Agronomy and Veterinary Institute, Morocco
ICAR	Indian Council for Agricultural Research
ICCROM	International Center for the Study of the Preservation and Restoration of Cultural Property
IDRC	International Development Research Centre
IFAD	International Fund for Agricultural Development
IFOAM	International Federation of Organic Agriculture Movements
ILRI	International Livestock Research Institute

IFT	Index of Frequency of Application (*French: Indicateur de Fréquence de Traitements*)
INGC	National Institute of Cereal Crops .
IP	Intellectual Property
INRA	National Institute of Agronomic Research, France
IPHAN	National Institute of Historic and Artistic Heritage, Brazil (*Portuguese: Instituto do Patrimônio Histórico e Artístico Nacional*)
IRD	Institute of Development Research, France
IRESA	Institute of Agricultural Research and Higher Education
IS	Innovation Systems
ISDA	Symposium on Innovation and Sustainable Development in Agriculture and Food
ITPGRFA	International Treaty on Plant Genetic Resources for Food and Agriculture
IUCN	International Union for Conservation of Nature
LAS	Localized Agrifood Systems
LINK	Learning Innovation Knowledge
NAIP	National Agricultural Innovation Project
NEG	New Economic Geography
NGO	Non-Governmental Organization
OECD	Organization for Economic Cooperation and Development
PGRFA	Plant Genetic Resources for Food and Agriculture
PMCA	Participatory Market Chain Approach
PPPs	Public-Private Partnerships
PSDR	Programme 'On and For Regional Development' (*French: Pour et Sur le Développement Régional*)
RAI	Rural Agro-Industries
RIU	Research Into Use
SACCAR	Southern African Centre for Cooperation in Agricultural Research and Training
SRI	System of Rice Intensification
SSA-CP	Sub-Saharan Africa Challenge Programme
TRIPS	Trade-Related Aspects of Intellectual Property Rights
TTOs	Technology Transfer Offices
UNCTAD	United Nations Conference on Trade and Development
UNDP	United Nations Development Programme
UNESCO	United Nations Educational, Scientific and Cultural Organization
VCC	Village Coordination Committee
WECARD	West and Central African council for Agricultural Research and Development (CORAF)

Preface

Patrick Caron

There is no doubt that the issue of innovation and its processes and impacts is fundamentally important for research institutions, policymakers and society as a whole.

As far back as 1993, at the initiative of CIRAD's Rural Economics and Sociology Mission, three French research institutions, CIRAD, INRA and ORSTOM (now IRD) organized an international seminar in Montpellier, France, on the topic of 'Innovation and Societies: What kinds of agriculture? What kinds of innovation?'. The 1990s were, in general, marked by a mobilization of the human and social sciences within agriculture research institutions. This was done in order to better understand and overcome what was perceived as social and cultural resistance to change and thus to facilitate and accelerate the dissemination of research outcomes. At the same time, representatives of these disciplines experienced firsthand the uncomfortable asymmetry of thoughts and actions between the dominant life sciences and the social sciences. Therefore this seminar was a welcome opportunity to take stock of innovation processes hiding behind these resistances and to encourage dialogue between disciplines.

Since then, in a context where environmental and poverty issues have gained importance, international agricultural research has shifted its focus to contributing to sustainable development and the Millennium Development Goals. This is reason enough to revisit once again the debates of the past. How best to position ourselves in a perspective of change and disruption?

What were our reasons to revive the debate, 17 years on, through another international conference? Our initiative was based on a four-fold motivation. First of all, we thought it important to base ourselves within a perspective of international thinking by broadening participation beyond only French institutions. Secondly, we recognized the real challenge of reaching beyond the conceptual framework of the human and social sciences and of seeking useful interdisciplinary fertilizations. Thirdly, we understood the challenge of looking beyond local dimensions – while recognizing the irreducibility of some local aspects – to address the interplay between different organizational levels, a challenge which echoes the global aspects of issues of development. Finally, meeting the actors of development and training, outside the confines of research, could help us better position ourselves in relation to changes that are taking place in society.

Research and education find themselves disrupted by technical, economic and social changes and the rise of uncertainties highlighted daily by the media. Environmental, economic, financial, social and political crises have become, in many cases, the basis for discussion and action. Such crises stress the urgent need for processes of adaptation and regulation based on collective action and call into question existing knowledge and challenge the omnipotence of scientific expertise. Innovation has always been perceived as a process where the solutions of tomorrow will be born. But the proven tendency of innovation in some cases to lead to situations of exclusion does not conform to the rhetoric and raises the question of the relationship between innovation and equity, highlighted in the IAASTD Report or the World Bank report (2008).

Current changes challenge in a very fundamental way the development models we consciously or unconsciously espouse and our unshakeable belief in the virtues of differentiation and innovation. Are we not systematically confusing innovation for development? Aren't our programmes still oriented towards change as a paradigm of social progress? Isn't innovation, as approached through the invention-innovation pairing, the very basis of action-oriented research? We can ill ignore these questions when we decide to focus on innovation.

It thus seemed appropriate to us to revisit our development models through a reflection on the link between Innovation and Sustainable Development, and this via an analysis of the effects of innovation processes in terms of 'development'. The scientist cannot let himself be trapped by an irrational belief in the virtues of innovation. Already in 1993, Chauveau and Yung[1], drawing on Hirshman, had counterposed two visions of these processes. On the one hand, a rhetoric, often called progressive or even scientific, supported by a diffusionist conception of innovation, defines innovation as a factor of technical and social progress. On the other, a rhetoric, called reactionary, critical of innovation, denounces the capturing of its benefits by dominant forces and challenges its idealized effects on traditional societies. This latter viewpoint is still very relevant and useful.

Thus, the objective and challenge of the ISDA conference on Innovation and Sustainable Development, organized jointly by CIRAD, INRA and Montpellier SupAgro, was to develop renewed, combined and cross-disciplinary perspectives on innovation. The intent was to understand the relationship between knowledge production, learning and innovation in terms of that ultimate of goals: development. The conference invited us to approach innovation itself as a research subject and to provide an update on scientific advances in this domain. It helped stimulate the reflexivity of our R&D activities and generate new ways of thinking about innovation. And, last but not the least, it encouraged innovation within the scientific world itself.

The conference was indeed an occasion for several fruitful discussions between the various communities working on innovation. With over 500 participants from 65 countries, it was a significant step forward in understanding innovation and its effects. I am convinced that the outcome of the conference will help shape our institutions' activities and programmes for some time to come. This book bears witness to the richness of debates and opens up new frontiers for our research institutions in the design of development policies.

[1] Chauveau, J.P. and Yung, J.M. (eds.), 1995. Innovation et Sociétés – V2 Les diversités de l'Innovation. Actes du XIVième séminaire d'économie rurale. INRA-CIRAD-ORSTOM, 13-16 September 1993, CIRAD, Montpellier, France.

Chapter 1. Reconsidering innovation to address sustainable development

Guy Faure, Emilie Coudel, Christophe T. Soulard and Hubert Devautour

1.1 Introduction

The world is being confronted by a multi-faceted systemic crisis. In addition to structural and ongoing changes such as climate change, increased pressure on renewable resources and population growth (still strong in Africa), the world must now contend with a severe economic crisis of unpredictable consequences, deepening poverty, shrinking export markets, tighter credit and cutbacks in development funding. In such a context, agriculture faces an uncertain future, particularly in some of the world's regions, with the emergence of differentiated development models that have led to an increasingly fragile family agriculture and the simultaneous rise of a capitalist agriculture. Nevertheless, this systemic crisis may also provide new opportunities over the long term. It is leading to a break from the past and calling into question paradigms until now taken for granted. It has brought to the fore the vulnerability of agricultural and agrifood systems and highlighted the need for innovation to take advantage of new development models. At a time of great uncertainty, with shifting values and standards, our societies should show themselves to be creative by reinventing modes of production, processing and distribution of agricultural products with a long-term perspective that takes into account the territories and their peoples, putting the concept of sustainability at centre stage.

Several agricultural and agrifood systems have already proven their ability to promote sustainable development by basing themselves on principles of agroecological production (De Schutter, 2011) or by encouraging local food systems (Muchnik and De Sainte-Marie, 2010). These innovative systems are still not very common and currently exist either in competition or in complement to the dominant productivist systems, some of them also claiming adherence to principles of environmental sustainability. These innovative agricultural systems take different forms depending on whether they are located in countries where agriculture is highly capital-intensive with high consumption of inputs and fossil fuels or in countries where agriculture has little access to these resources, resulting in low labour productivity.

While it is necessary to report on, share and capitalize on these innovative experiments, it is more urgent to create a new paradigm to consider differently the contributions of the agricultural sector to development. A renewal of agricultural and agrifood systems is not accomplished by simply designing new technical and/or organizational solutions. It has to examine the very status of knowledge necessary for sustainable development and call into question the monopoly of scientific knowledge over other forms of knowledge. It requires transforming the innovation processes by creating new links between research, economic stakeholders, civil society actors and policymakers. Agricultural research should reorient itself in its involvement with innovation, given that agriculture is no longer assessed by its sole function of production but rather by how this production interfaces with the environment and society as a whole. The issues to address today are 'agriculture and health', 'agriculture and environment', 'agriculture and energy', 'agriculture and rural activities', etc. To understand these changes and their implications for research, we must engage in collective thinking.

Various recent works have proposed new directions for encouraging innovation. Thus, the books 'Innovation Africa' (Sanginga *et al.*, 2009) and 'Farmer First Revisited' (Scoones and Thompson, 2009) stress the need to promote collaboration between farmers, researchers, advisory services and the private sector to create useful practical knowledge and to improve technologies by adapting them to farmer requirements. The role of the market as a driver of innovations is becoming increasingly accepted. Emphasis is laid on capacity building of actors and thus on training and advice, on strengthening social capital and thus networks, on the establishment of new institutional arrangements including the promotion of multi-actor platforms and on drawing up of suitable public policies. The book 'Action Research in Partnership' by Faure *et al.* (2010) focuses on new ways of conducting research by according greater importance to interactions between stakeholders.

These different options are part of a lively debate with contrasting positions being espoused on the type of innovation necessary for sustainable development. That is why we wanted to initiate a clash of ideas and sharing of thoughts on this topic through an international symposium, 'Innovation and Sustainable Development in Agriculture and Food' (ISDA), which was held in June 2010, in Montpellier, France. Participants included researchers from various disciplines, development actors and policy makers from countries of the North and the South. This book presents an account of these reflections, which analyzed experiences undertaken to promote innovation, drew lessons from their successes and failures, with the hope that they will lead to the emergence of new scientific and political perspectives for innovation systems that could contribute to sustainable development.

The chapters of this book express various positions on innovation processes and contain reflections of different experts in the dynamics of innovation, of institutional representatives involved in guiding and managing innovation, and of researchers who have analyzed and participated in innovations on the ground. The authors have relied on the many presentations and discussions held during the ISDA symposium; some of them will illustrate their points in the text with case studies[2].

[2] The symposium proceedings are available online: http://hal.archives-ouvertes.fr/ISDA2010.

1.2 Innovation for what kind of development?

1.2.1 Development, a constant questioning

To understand innovation and its contributions to development, it is important to consider the meaning given to the notion of development by various actors. Indeed, the definition of development is not self-evident. A first definition, widely disputed but which still surfaces in current debates, argues that *'development is economic growth'*, as measured primarily by a nation's gross domestic product. Further refinements are possible by incorporating other criteria such as social justice: *'development is evenly distributed economic growth'*. Economic thinking on development has been marked by the necessary evolution of societies due to growth. For example, Rostow (1979) identifies five stages of growth through which societies must pass to get closer to 'developed' Western societies: (1) the traditional agricultural society, (2) expansion of trade with the first changes in techniques and attitudes, (3) the 'take-off' powered by cumulative growth, (4) the 'drive towards maturity' with progress extending to all activities and (5) the advent of a mass consumer society. Such a definition of development is marked by the notion of progress with a clearly defined target, shared by a large section of society and towards which society proceeds step by step. Its adoption implies a period of major planning of technical interventions and a belief in the overall positive effects of these techniques.

As far back as the 1970s, 'The Club of Rome' warned of the danger of unfettered economic and population growth leading to the depletion of natural resources. The development model based on the accumulation of wealth was then contested. François Perroux (1963) proposed a more social definition: *'development is a combination of a population's mental and social changes which makes it ready for cumulative and durable growth of its real global product.'* One of the proposed alternatives to economic development was that of Maurice Strong, secretary general of the 1972 Stockholm conference. He spoke of *'eco-development'*, in the sense of a prudent use of resources and a valorisation of Third-World knowledge. This concept, which was also taken up by Ignacy Sachs (1980), became central to the policies of the United Nations programmes of the 1980s. At the same time, the future Nobel laureate in economics, Amartya Sen (1989, 1999), proposed a relook at poverty through the prism of basic needs and capabilities of individuals and freedoms enjoyed by them. He introduced the concept of human development (later formalized as the HDI, the human development indicator). During the Rio conference in 1992, these ideas all converged together in the notion of sustainable development as enunciated by the famous definition by Brundtland: *'sustainable development is development that meets present needs without compromising the ability of future generations to meet their own needs'* ('Our Common Future' report of the World Commission on Environment and Development, 1987). This new paradigm emphasized the multiplicities of the dimensions of development with the ones most often mentioned being economic, social and environmental. Cultural and governance-related dimensions, are also widely cited.

Thus development becomes the implementation of a social project. The nature of the project is not defined *a priori* in reference to one or other external models. Its objectives can differ from one society to the next and there are several paths to reach them. What counts is the flexibility, the resilience and the reversibility of economic and social systems put in place to avoid compromising the ability of societies to ensure their futures.

1.2.2 Development in agriculture

The agriculture sector is, of course, impacted by this thinking on development. For decades, agriculture has been driven by the goal of increasing production by, on the one hand, promoting an increase in cultivated area per unit of labour or capital and, on the other, by raising crop and livestock yields. This model was called the 'green revolution' in countries of the South and intensive agriculture in those of the North. It was only gradually that other dimensions relating to territories, sectors, value chains or food systems were taken into account, at more or less the same time that limitations of positivist and productivist models were becoming apparent. The concept of multifunctionality of agriculture that emerged in the 1990s is one example. It recognizes that agriculture, beyond its function of production, plays an important role in building a territory, participates in the management of renewable resources, generates jobs and helps build a local culture. Thus, gradually, other agricultural models emerged, based on novel principles such as agroecology (Altieri, 2002) or seeking to promote new short supply chains or new forms of equity in the markets (Colonna *et al.*, 2011).

What strikes the observer is how these new agricultural models, although well thought out in theory and gradually being implemented on the ground, are unable to displace the old models. Instead, different models coexist on the same territory. In Brazil, for example, family farming, State-supported for the past several decades, coexists with a capitalist agriculture where the role of finance is gradually gaining importance. In Europe, the model based on the family farm can no longer form the basis of agricultural policy given the appearance of private investment in the farming sector and the growing importance of diversification of activities of rural and agricultural households. In Africa, the phenomenon of land-grabbing is the brutal materialization of these rapid changes, with large private enterprises taking priority over traditional communities in matters of land ownership. While the coexistence of these models in a territory may be observable, it is far from easy to assess their dynamics and respective contributions. Are there conflicts between the models, especially for access to resources? What synergies exist, for example, in the creation of new markets? This coexistence can also be found even within organizations, for example, with a farmer association selling some of its products in a niche market and others through the mass-distribution route. It is also evident at the level of consumers who can purchase a product labelled 'fair trade' at a farmers' market or cooperative shop and another cheap product the same day at a discount supermarket.

The diversity of models calls into question the place of farmers in agricultural development. In a meaning close to Sen's (1999) 'capability', development can correspond to the capacity building of farmers so that they can, on the one hand, define their own targets and, on the other, acquire and implement the means of achieving them. Capacity building also helps them increase their independence and makes them more self-reliant. However, as stated, this concept is not completely unambiguous. Indeed, if farmer collectives can shape and be part of such a process, there is no guarantee that they will contribute to the building of assets or benefits considered as commons by the rest of society. We find here a 'negotiated' idea of development where the definition of what agriculture should be involves diverse agricultural and non-agricultural stakeholders with differing requirements (Compagnone *et al.*, 2009).

1.2.3 The place of innovation in development

Innovation is at the core of reflections on development, most often simply seen as the engine of development. Schumpeter (1934) was the first to propose that innovation allowed an entrepreneur to acquire a comparative advantage over his competitors and thus to generate profits. According to him, innovation is a new combination of production factors which can be expressed by the making of a new product, the devising of a new production method, the building of new outlets and markets or access to new resources. Innovation is an invention that has found its market and has thus become part of a system of production. Subsequently, numerous studies have attempted to characterize innovation which, for example, can be incremental or radical, technical or process-oriented, driven by the market (pull innovation) or by technological advances (push innovation). It is traditional to say that innovation can take various forms: technical, economic, organizational, social, etc. More specifically, innovation is usually composite: a technical innovation is most often paired with an organizational innovation in which we could say it is embedded. This observation leads to the concept of the socio-technical innovation[3], with the technical object being understood through the uses it is put to and the social ties that its use generates, modifies or destroys (Flichy, 1995). Innovation can be grafted onto older systems or constitute a break from the past, it can be exogenous in origin and be driven by the technicians or endogenous with the farming world as its source. It can emerge in very varied contexts and can only be understood by an analysis of the overall society and the context in which it develops. Every innovation has a history: it is born, it develops and then it dies. Innovations are thus strategic instruments to achieve the objectives of certain actors or bolster their positions. They can be used as tools of power and negotiation by certain sectors of society to press for their preferred development agendas.

At another scale and in modern times, Europe claims to see its development solely in a knowledge and innovation society in order to maintain the competitiveness of countries and thus their societies' prosperity. Innovation can be found on the agenda of several

[3] The socio-technical analysis considers an object in its social milieu and places itself at the exact point where the innovation is located (Akrich *et al.*, 1988).

stakeholders (policymakers, business leaders, research institutions, etc.) because it fits well in a context of production that is evermore complex and uncertain, where the combination of physical, social, intellectual dimensions of change enable continuous adaptation to an unstable environment which is in turn being constantly modified by this combination. But if innovation is not just technological but refers usually to complex processes, it does not mean that every change within an organization, or even a society, is an innovation. Innovation is a process that intentionally promotes the emergence of the new and of its adoption by society. Innovation requires a project of change and an interaction between the actors promoting this change and those appropriating it. Olivier de Sardan (1993) illustrates the complexity of the phenomenon through his definition of innovation: *'a novel graft, between two indistinct groups, in an arena, via brokers.'*

Innovation thus directly questions the development models that we promote, consciously or unconsciously. Yung and Chauveau (1993) distinguish a progressive and scientific rhetoric of innovation as a source of technical and social progress in a diffusionist view from a reactionary rhetoric critical of innovation which denounces the annexation of benefits by dominant players to the detriment of traditional social structures. The central issue is not the improvements in performance of actors involved in innovation but rather the consequences of innovation on the other actors in terms of their inclusion or exclusion. Aimed at improving land and labour productivity, technical progress promoted in the agricultural sector has undermined the future of small farmers everywhere (Röling, 2009).

1.3 Research on innovation changes to progressively take into account complexity

Research has always played an important role in understanding and promoting innovation. Is it not the very foundation of its activities? Nevertheless, research practices have evolved to better address the increasingly complex problems confronting societies. This evolution is based on research gradually (and partially) calling into question its ways of producing knowledge and of taking into account the 'social' requirements (which actors, what goals, what consequences?), leading to a different way of perceiving innovation. Innovation has thus been successively characterized by the behaviour of individuals, by interactions between individuals (networks) and, later, by interactions between organizations. We track here this evolution, in particular for the agricultural and agrifood domains.

1.3.1 The farmer's role in the innovation process

Until the 1980s, research focused on the producer and his environment for analyzing innovation. Rogers (1983) showed the different attitudes and behaviours of producers when confronted by change and drew up a typology of innovators (venturesome innovators, early adopters and social leaders, early majority, traditional late majority, wait-and-see laggards). He showed that the dissemination of innovation is non-linear, following an S-curve with time on the x-axis and the number of adopters on the y-axis. But this diffusionist model does not take into account the fact that the conditions relating to the farm or its environment, which can either encourage or discourage the innovation under consideration, are not the same for all. As emphasized by Olivier de Sardan (1993), the diffusion of innovation, especially in the early stages, depends also on its proponents and on their ability to promote it based on their social position and the interests they perceive. For their part, Mendras and Forsé (1983) proposed five factors to evaluate the 'adoptability' of innovations, namely: the relative advantage accorded by the innovation compared to the initial situation, its compatibility with the system already in place, its greater or lesser complexity, its 'trialability' in the context of the actor concerned, and the observability its results with others. These factors incorporate the innovation's degree of complexity and the level of risk that the producer is being asked to take. But the question doesn't really have to do with the compatibility with the existing production system or if the producer can adapt to the proposed innovation. The real issue is to understand the dual relationship that is being established: how does the innovation change the farm and how does the farmer incorporate the innovation by adapting it?

Chauveau *et al.* (1999) relativizes this way of characterizing innovation, which puts the focus on the individual by emphasizing the interactions between him and his environment. Thus, they specify that (1) the supply and demand of innovation are built during interactions between the actors on technical issues, (2) innovations propagate through composite networks taking the heterogeneity of socio-economic units into account and (3) the relationships between innovation and the economic, social and political environment are not linear processes.

1.3.2 The diffusionist model called into question

Based this body of work, the ability of the dominant linear model of knowledge and technology transfer, called the diffusionist model, to provide answers to complex problems is being called into question. Nevertheless, it is still very much alive; it is not rare for researchers of the natural sciences to seek support from the human or social sciences to disseminate technologies developed in the laboratory or at the field station. However, debate on the changes necessary in the positioning of research in innovation processes remains lively. The story of research in developing countries is illustrative of the changes observed. From the 1980s, the research-development approaches to improving farm performance while taking rural realities into account have resulted in extensive literature.

Jouve and Mercoiret (1987) define research-development as *'full-scale experimentation in close consultation with the farmers for the technical, economic and social improvement of their production systems within their environment'*. It seeks to create a reciprocal triangular relationship between research, extension services and the farmers at every stage of the process of transforming production conditions. French research-development approaches and Anglo-Saxon ones of 'farming system research' converge on the basics (Jouve, 1994): willingness to consider the actual production conditions, a systemic approach to complex agrarian realities, conceptualization of the farm as a system, analysis of farmer practices, farmer participation in the processes of farm improvement and the need for a political and social environment conducive to innovation.

Research-development has been subjected to strong criticism for the high cost in money and time of its diagnostic phases, the weakness of its recommendations in terms of useful action and the risk of exploitation of producers during the testing process. This criticism has led to changes aimed at encouraging a deepening of research-development approaches for an individual and collective learning process. Research in the English-speaking world has also evolved. Criticism of the 'farming system research' projects of the 1970s and -80s (Norman and Collinson, 1985) was followed by accelerated methods of participatory research (Chambers *et al.*, 1989; Scoones *et al.*, 1994) which successively emphasized 'rapid rural appraisal', 'participatory rural appraisal' and, more recently, 'participatory learning action', as shown by Lavigne Delville *et al.* (2000).

1.3.3 Innovation networks

The debate on innovation has thus shifted gradually to the analysis of interactions between individuals who are part of a network or work within organizations. The concept of social capital characterizes the networks formed between the actors and, in particular, the type, density and intensity of their relationships. This approach considers that an individual's or group's development potential, and thus the ability to innovate, is related to his or its social capital (Coleman, 1988). In research on innovation, the emphasis is increasingly placed not only on the learning processes at the individual level but also at the group and organizational levels in order to produce knowledge that can be useful in bringing about a desired change. For the organizational context, Argyris (1995) developed theories of single-loop learning (change of practices) and double-loop learning (change in values). The concept of community of practice (Wenger, 1998) helps understand how individuals, who frequently share ideas on a given issue, learn in their professional activities and generate new knowledge validated by this community.

In the agricultural domain, Darré (1996) shows that farmer networks of dialogue and work are the source of knowledge creation. These networks consist of clusters of individuals or 'local professional groups' having regular meetings and discussions on topics related to their farming activities. A new standard is discussed within the group through a dialogue that is more or less intensive depending on circumstances. The group then modifies the standard, accepts it or rejects it. The group thus produces knowledge about what to do and how to do it. But not every member has the same status in the group; some are more respected than other. Thus every argument is assessed by each group member in two dimensions: the argument's intrinsic value corresponding to its interest for action according to the norms of the group and the argument's social value corresponding to the status of the individual who advanced it.

When actor groups are not homogeneous or structured within an organization, Akrich *et al.* (1988a,b) show that innovation only becomes reality if it is a stake for the actors concerned and if they can incorporate it in their own strategies. Innovation is the art of making an increasing number of allies to become interested, thus strengthening the process. Not only its contents but its very chances of success lie in the choice of representatives or spokespersons who will interact and negotiate to give shape to the project and modify it as necessary for successful realization. Nevertheless, innovation is far from being a planned (or plannable) linear process; it resembles rather a whirlwind model with multiple decision makers, with a myriad of small and large decisions, much uncertainty and a lot of apparent confusion. Olivier de Sardan (1998) emphasizes the social proponents who have a more or less recognized status in local society and through whom the innovation has to pass. Similarly, innovation usually induces deferred direct and indirect effects on the local social structure by serving certain interests and thwarting others. It can thus consolidate the existing social structure or, on the other hand, promote social change.

1.3.4 Innovation systems

To explain networks that form between organizations, some authors (Röling, 1990; Engel and Salomon, 1997) go beyond both the linear adoption model and the informal networks and formalize the existence of 'knowledge and information farming systems' where research and educational institutions and agricultural advisory services are perceived as so many actors amongst the many others in the dynamics of innovation. Research thus takes place in territorial or sectoral innovation systems.

The role of organizations in the innovation processes was first formalized in the industrial world. Within a same territory, firms and service providers can form a cluster (Porter, 1990) in a coherent social whole, sharing common histories and values and recognized for its innovation dynamics. At the country level, research and educational organizations responsible for producing knowledge and industrial actors involved in innovation can form a national innovation system (Freeman, 1988; Lundvall, 1992). In this system, the actors interact and create new knowledge, know-how and expertise. A consequence is that relationships between science and society are often transformed. According to Gibbons and his collaborative working group (1994), this transformation is a break with the traditional view of State, industry and science as separate and distinct spheres. Nowotny *et al.* (2001) also emphasize the fact that expertise is now 'socially distributed', which leads to the conclusion that everyone contributes to knowledge, i.e. 'we are all experts'. However, this perception of the merging of the spheres of knowledge is not shared by all analysts. Etzkowitz and Leydesdorff (2000) have proposed another analytical model of the science-society relationship where three helices represent, respectively, the spheres of public administration, science and industry. This model takes into account the interaction between these spheres and their co-evolution through institutional, technological and scientific transformations. The development of lateral relations between institutional spheres encourages the formation of new areas of knowledge, while allowing changes to take place within each sphere.

The triple helix model is however criticized by Shinn (2000) because it does not take the existence of a diversity of regimes of knowledge into account, i.e. of a plurality of modalities to link scientific and socio-economic actors. The organizations are indeed not all equal and do not have the same innovation capabilities: those incapable of modifying their routines to new limitations or take advantage of new opportunities eventually disappear as shown by evolutionary theory (Dosi *et al.*, 2000). This approach can take technological trajectories (Dosi, 1982) and the evolution of innovation regimes into account. Dominant regimes may experience lock-ins for institutional reasons and alternative regimes may replace an existing dominant regime or co-exist or hybridize with it.

The complex interplay between actors – individual ones or organizational ones – has led to the introduction of the concept of 'innovation systems' for an improved understanding of innovation dynamics. An innovation system can be defined as *'a network of organizations, enterprises, and individuals focused on bringing new products, new processes, and new forms of organization into economic use, together with the institutions and policies that affect their behaviour and performance'* (Rajalahti *et al.*, 2008).

An innovation systems approach needs to address several important issues:

- The limits of the system are not provided but are built: they depend on the type of questions asked and on the point of view of the persons doing the asking (Carlson et al., 2002). Thus, the literature has analyzed local, sectoral, national, even international innovation systems.
- Following the limits set for the system, it is not always easy to identify the actors and characterizing their resources (material and financial, of course, but also in terms of knowledge, skills and abilities) may often be difficult.
- More than the actors present, it is the nature of their interactions that allows the innovation system to be characterized by specifying the nature of the networks (formalization, density, flexibility, etc.). This permits access to resources, generation of knowledge and promotion of individual and collective learning processes (Spielman *et al.*, 2009). In these networks, some actors have enhanced statuses and act as intermediaries (Klerkx and Leeuwis, 2008).
- The nature of technology at play in the innovation system (incremental or radical innovations, 'product' or 'process' innovations) has an influence on the scope and nature of the innovation system.
- Governance of the innovation system refers to the relationships between actors and formal and informal mechanisms established to plan actions and monitor and assess results. This governance also depends on the type of innovation: driven by the market or by upstream or downstream production operators; promoted by the producers or communities to seize an opportunity or overcome a limitation; or facilitated by entrepreneurs.

- Public policies that can enhance the innovation systems' performance, in particular by influencing support mechanisms for innovation (platform, network, specialized institution) and for professional training, using funding from both the private and public sectors.
- Evaluation of the performance and impact of innovation systems, which raises the question of the criteria used to do so and thus of the goals and the models aimed for.

1.4 New questions, new debates

This brief history of the work on innovation shows a clear progression of scientific thought. At the start, the individual took the dominant role in explanations of the processes of innovation even if his role was put in a broader context and if the conditions that promote or inhibit innovation were taken into account. Current research relativises the weight of individual behaviours to focus on interactions between individuals within social networks, between individuals within organizations or between organizations. The field of thinking on innovation has expanded to encompass work on organizations (those concerned with agricultural advisory services, research, training or even funding) and on institutions that define the playing field and rules for these actors. This system-based approach helps explain the diversity of situations using interactions between different elements of the system at various levels (local, regional, national, sectoral), even if not much work exists on the links between these different levels and thus on issues specific to each level. For some systems, it is the tension between the actors that directs the innovation process and often leads to highly unstable or blocked situations. For others, it is the construction of shared goals that prevails, often leading to one or more collective actions, which can explain the system's evolution over time.

The system-based approach to innovation thus questions the meaning assigned to innovation. Described in early research as new technology, a new production or marketing method (Schumpeter, 1934), innovation is now understood as the result of complex interactions, involving individuals and organizations acting in multiform networks, which can lead to synergies or reveal opposition. All this work highlights the complexity and non-linearity of innovation (from a designer to the end-user or from initial stage to final stage) and thus the inability to anticipate or forecast innovation even if some actors can stimulate the process. In a context where development models (even sustainable ones) are fiercely debated and involve important societal choices, innovation appears clouded with uncertainty. This difficulty raises the question of the ability of actors to innovate in order to create new production systems and activities for sustainable agriculture in different regional contexts.

Does the emergence of the concept of innovation systems lead to a better debate on the different development models? What are the current challenges confronting innovation systems? The first part of this book invites us to examine these questions in depth from three different perspectives.

Lawrence Busch (Chapter 2) provides us with critical reflections of American sociology on the role of standards and the growing influence of the private sector. He shows how standards to assess agricultural performance, existing since its very beginnings, guide the type of research undertaken. He calls for reflection on the standards that direct current agricultural research, now increasingly becoming a private property of large companies and limiting/freezing/locking-in agricultural innovation on the ground. How, in such a context, to return to current challenges: food security, inequality and climate change? It is time to develop new standards to evaluate research outcomes and go beyond the sole criterion of productivity.

Rikka Rajalahti and Bernard Triomphe present the thinking developed by the World Bank on innovation systems, in an initiative involving different stakeholder categories, including research institutions. Presented as a new way of looking at innovation, involving many actors beyond those of formal research, innovation systems pose challenges in terms of establishing policies and institutional mechanisms to allow their implementation on the ground. Various sessions and a roundtable organized on this topic at ISDA were successful in identifying current challenges, which are presented in Chapter 3.

The concept of the innovation system may indeed allow one to reconsider the coordination of innovation but there is the risk of overlooking the very many innovative experiences that are thriving under the radar of research. This is the message of Andy Hall and Kumuda Dorai (Chapter 4) based on the LINK network (Learning INnovation Knowledge) that they have been managing actively for the past several years. Using various examples, they show us the importance of identifying innovative entrepreneurs, often invisible even though they are behind many dynamics of change. These authors suggest paths for research to better support such entrepreneurs and encourage them in their pioneering work.

Following these invitations to reconsider innovation differently, the second part of the book addresses practical on-the-ground questions of innovation for sustainable development. These questions were the basis of the ISDA symposium's sessions: How to reconcile agricultural production and environmental preservation? What are the consequences of exclusion from innovation? How to approach innovation from a perspective of a more equitable development? How to be creative and generate new knowledge useful for innovation? How to organize collectively at the level of production chains or territories in order to promote innovation? The authors of each of the chapters in this part of the book were given the responsibility of summarizing the original contributions made in the ISDA presentations, while relying on their research experience in the analysis and support of innovation.

Jean-Marc Meynard presents an agronomic viewpoint by exploring, in Chapter 5, the effects of the intensive agricultural model as used for specialized cereal crops. Going beyond just the negative environmental consequences, he shows the systemic effects that reinforce the pursuit of this innovation regime and the lock-in effects induced by excessive specialization, with actors finding themselves unable to redeploy their skills to other types of development models. When thus trapped, how can the actors free themselves? While warning that there are no 'good solutions,' Meynard proposes the basics of an 'agro-ecological engineering' approach which would accord the actors the flexibility necessary to move freely and confidently into the future.

Denis Requier-Desjardins (Chapter 6) examines the ability of innovation to reduce the vulnerability of the poorest segments of rural populations. According to him, when the fight against poverty is taken up in terms of the building of actor 'capabilities,' i.e. by focusing on their ability to make life choices and find their place in society, agricultural innovation no longer remains as central a concern as before. Indeed, the diversity of activities (agricultural or non-agricultural), immigration and integration into marketing channels within the social economy can be viewed as innovative strategies to escape poverty. While new policies of poverty alleviation do indeed seek to strengthen human capital, they rarely encourage these various forms of social innovation, warns Requier-Desjardins.

In Chapter 7, Estelle Bienabe, Johann Kirsten and Cerkia Bramley explore innovations driven by agrifood markets. These innovations take the form of certification of 'sustainable' standards. These standards need therefore to be defined. The authors examine, in particular, the types of collective actions or social movements at the origin of these certifications and their consequences in terms of fairness and inclusion of small farmers. But not everyone has the same ability to leverage the benefits of these certifications nor the perseverance necessary to maintain their long-term participation in the supply chains that are created. To avoid the exclusion of small producers, the authors stress the importance of a real social construction of 'quality' through the establishment of empowerment processes and through the definition of private and public regulations able to structure the development of these new standards.

André Torre and Frédéric Wallet, in charge of the research programme On and For Regional Development[4] in France, begin Chapter 8 with the observation that regional studies often emphasize a vision of territories as 'development poles' constructed around high-tech innovations. But this race for technological excellence is quite illusory in the case of many rural territories. What then are the paths of innovation for these territories? By analyzing various types of innovation policies, the authors examine ways of constructing territorial innovations that are truly based on the strengths of a rural territory and on the 'wishes' expressed by its population. It is through territorial governance, seen as a space for dialogue and for expressing opposition, that multi-level, multi-actor and multi-faceted innovation processes can emerge and grow.

[4] *Pour et Sur le Développement Régional* (PSDR) in French: This regional development programme brings together regional partners and research, in ten French regions, in direct contact with the actors and their concerns and with equal funding by the regions and research organizations.

These perspectives of on-the-ground observations show that there are interesting localized experiments which can form the basis for development. But it still remains difficult to operationalise the concept of the innovation system to propose, and especially prioritise, public policies for stimulating and strengthening innovation processes at the scale of a territory or economic sector. There exist several options to do so: build individual capacities through training, in particular on the topic of entrepreneurship; structure organizations dedicated to innovation, especially certain research centres; and foster positive synergies between actors such as, for example, private-public partnerships between research institutions, training centres and companies. But it is undeniably more important to act on the governance itself of innovation systems through better identification and strengthening of where and how innovation processes are oriented and assessed. Also strategic is the issue of funding for innovation, both for rationalizing public investments as well as to facilitate access to capital for the different types of actors seeking to innovate.

To shed light on these implications for public policies, we wanted to give the floor to representatives from institutions from different continents so that they could share their perceptions and visions. What can policy makers do to encourage new ways of innovating? How to institutionalize and consolidate alternative experiences into true large-scale innovation systems? What new policy tools can and need to be developed?

In Chapter 9, Juliana Santilli, Public Prosecutor at the Public Ministry of Brazil, shows, in the context of preservation of biodiversity, that international institutions propose paths of legal innovation that States can translate into laws. Brazil has, in this manner, enacted laws on the protection of agrobiodiversity, defying conventional laws on seed ownership which have long benefitted multinational corporations. These new laws guarantee the right of family farmers to use and multiply their own seeds. Actual implementation of these laws is still in its infancy but this legal affirmation is a first step towards a complete reorganization of the national seed system.

Karim Hussein, of the OECD, and Khalid El Harizi, of IFAD, present in Chapter 10 the results of a roundtable involving different innovation-oriented policymakers from around the Mediterranean. They argue that simply designing policies for innovation is not sufficient; these policies must also form part of a broader institutional and political framework. Through examples of policies at different levels, ranging from the local to the international, they stress the importance of development of skills of different types of actors and highlight the need for platforms for exchanges in order to design more flexible policies, easily adaptable to the diversity of requirements of actors of innovation. Following the Arab Spring, these considerations call into question the ability of governments to allow real participation by local stakeholders in the processes of innovation.

In Chapter 11, Papa Seck, Director of Africa Rice, and his co-authors Aliou Diagne and Ibrahima Bamba, invite us to revitalize the governance of African agricultural research.

For them, this can only happen when researchers will be engage in more participatory approaches, which implies more decentralized decision-making and resource-management mechanisms. The co-construction, co-execution and co-evaluation of agricultural polices has become indispensable for an efficient agriculture. This cannot happen without a major cultural and organizational change.

In Chapter 12, the conclusion to the book, we put in perspective the issues addressed during the ISDA symposium. Even if new ways of thinking about innovation are gradually emerging, even if interesting experiences for innovating differently are taking place, it still appears that current development models and innovation policies are inadequate to meet the enormous challenges facing the agricultural world. Our main point in the conclusion is to question this observation. What can then be done to rethink innovation in the current context? How to deconstruct models that prevent us from approaching innovation differently? How to strike off in new directions and construct new knowledge in an uncertain environment? ISDA provides us with some answers but the issues remain open for consideration in future meetings.

References

Akrich, M., Callon, M. and Latour, B., 1988a. A quoi tient le succès des innovations. Premier épisode: l'art de l'intéressement. Deuxième épisode: l'art de choisir les bons porte-parole. Gérer et comprendre, Annales des Mines, 1988 (11), 4-17.

Akrich, M., Callon, M. and Latour, B., 1988b. A quoi tient le succès des innovations. Deuxième épisode: l'art de choisir les bons porte-parole. Gérer et comprendre, Annales des Mines, 1988 (12), 14-29.

Altieri, M., 2002. Agroecology: the science of natural resource management for poor farmers in marginal environment. Agriculture, Ecosystems and Environment, 93, 1-24.

Argyris, C., 1995. Savoir pour agir. Surmonter les obstacles à l'apprentissage organisationnel. Inter Editions, Paris, France.

Carlsson, B., Jacobsson, S., Holmén, M. and Rickne, A., 2002. Innovation systems: analytical and methodological issues. Policy Research, 31, 233-245.

Chambers, R., Pacey, A. and Thrupp, L.A., 1989. Farmer first. Farmer innovation and agricultural research, Intermediate Technology Publication, London, UK.

Chauveau, J.P., Cormier Salem, M.C. and Mollard, E. (eds.), 1999. L'innovation en Agriculture, questions de méthodes et terrains d'observation. IRD, Paris, France.

Coleman, J., 1988. Social capital in the creation of human capital. American Journal of Sociology, 94, 95-120.

Colonna, P., Fournier, S. and Touzard, J.M., 2011. Systèmes Alimentaires. In: Esnouf, C., Russel, M. and Bricas, N. (eds.) 'Pour une alimentation durable: réflexion stratégique duALIne'. Editions Quae, Versailles, France, pp. 79-108.

Compagnone, C., Auricoste, C. and Lémery, B. (eds.), 2009. Conseil et développement en agriculture: quelles nouvelles pratiques? Educagri et éditions Quae, Dijon, France.

Darré, J.P., 1996. L'invention des pratiques dans l'agriculture. Karthala, Paris, France.

De Schutter, O., 2011. Rapport du Rapporteur spécial sur le droit à l'alimentation. Assemblée Générale des Nations Unies, Geneve, Switzerland.

Dosi, G., 1982. Technological paradigms and technological trajectories: a suggested interpretation of the determinants and directions of technical change. Research Policy, 11, 147-162.

Dosi, G., Nelson, R.R., and Winter S.G. (eds.), 2000. The nature and dynamics of organizational capabilities. Oxford University Press, New York, NY, USA.

Engel, P.G.H. and Salomon, M.L., 1997. Facilitating innovation for development. A RAAKS resource box. KIT Publishers, Amsterdam, the Netherlands.

Etzkowitz, H. and Leydesdorff, L., 2000. The dynamics of innovation: from national systems and 'Mode 2' to a triple Helix of university – industry – government relations. Research Policy, 29, 109-123.

Faure, G., Gasselin, P., Triomphe, B. and Temple, L., (eds.), 2010. Innover avec les acteurs du monde rural: la recherche-action en partenariat. Quae, CTA, Gembloux, France.

Flichy P., 1995. L'innovation technique: récents développements en sciences sociales. Vers une nouvelle théorie de l'innovation. La Découverte, Paris, France.

Freeman, C., 1988. Japan: a new national system of innovation? Technical Change and Economic Theory. Pinter, London, UK.

Gibbons, M., Limoges, C., Nowotny, H., Schwartzman, S., Scott, P. and Trow, M., 1994. The New Production of Knowledge. SAGE Publication, Thousand Oaks, CA, UK.

Jouve, P., 1994. Le diagnostic des conditions et modes d'exploitation agricoles du milieu. De la région à la parcelle. In: L'appui aux producteurs ruraux. Ministère de la Coopération-Karthala, Paris, France, pp. 57-98.

Jouve, P. and Mercoiret, M.R., 1987. La recherche-développement: une démarche pour mettre les recherches sur les sytèmes de production au service du développement rural. Les Cahiers de Recherche-Développement, 16, 8-15.

Klerkx, L. and Leeuwis, C., 2008. Matching demand and supply in the agricultural knowledge infrastructure: experiences with innovation intermediaries. Food Policy, 33, 260-276.

Lavigne Delville, P., Sellamna, N.E. and Mathieu, M., 2000. Les enquêtes participatives en débat. Ambitions, pratiques et enjeux, Col. Economie et Développement, Karthala, Paris, France.

Lundvall, B.A., 1992. National systems of innovation. Pinter, London, UK.

Mendras, H. and Forsé, M., 1983. Le changement social. Armand Colin, Paris, France.

Muchnik, J. and De Sainte-Marie, C., 2010. Le temps des SYAL. QUAE Editions, Versailles, France.

Norman, D. and Collinson, M., 1985. Farming systems research in theory and practice. In: Agricultural Systems research for developing countries, Richmond, Australia, 12-15 May 1985, ACIAR, No. 11, pp. 16-30.

Nowotny, H., Scott, P. and Gibbons, M., 2001. Re-thinking science: knowledge and the public in an age of uncertainty. Polity Press, Cambridge, UK.

Olivier de Sardan, J.P., 1993. Une anthropologie de l'innovation est-elle possible. In: Séminaire d'Economie Rurale 'Innovation et sociétés', 13-16 September 1993, Montpellier, pp. 33-49.

Olivier de Sardan, J.P., 1998. Anthropologie et développement. Essai en socio-anthropologie du changement social, APAD-Karthala, Paris, France.

Perroux, F., 1963. Introduction à l'économie du XXème siècle. PUF, Paris, Fracne.

Porter, M.E., 1990. The competitive advantage of nations. MacMillan, Basingstoke, UK.

Rajalahti, R., Janssen, W. and Pehu, E., 2008. Agricultural Innovation Systems: from diagnostics toward operational practices. Discussion Paper 38, World Bank, Washington, DC, USA.

Rogers, E.M., 1983. Diffusion of innovations, 3rd ed. Free Press, New York, NY, USA.

Röling, N., 1990. The agricultural research-technology transfer interface: a knowledge systems perspective. In: Kaimowitz, D. (ed.) Making the link: agricultural research and technology transfer in developing countries. Westview Press, Boulder, CO, USA, pp. 1-4 and 11-23.

Röling, N., 2009. Conceptual and methodological developments in innovation. In: Sanginga, P., Waters-Bayer, A., Kaaria, S., Njuki, J. and Wettasinha, C. (eds.). Innovation Africa: enriching farmers' livelihoods. Earthscan, London, UK, pp. 9-34.

Rostow, W.W., 1979. Stages of economic growth: a non-communist manifesto. Cambridge University Press. Cambridge, UK.

Sachs, I., 1980. Strategies de l'ecodeveloppement. Ouvrières, Paris, France.

Sanginga, P., Waters-Bayer, A., Kaaria, S., Njuki, J. and Wettasinha, C. (eds.), 2009. Innovation Africa: enriching farmers' livelihoods. Earthscan, London, UK.

Schumpeter, J.A., 1934. The theory of economic development: An inquiry into profits, capital, credit, interest, and the business cycle. Harvard University Press, Cambridge, MA, USA.

Scoones, I., Thompson, J. and Chambers, R. (eds.), 1994. Beyond farmer first: rural people's knowledge, agricultural research and extension practice. Intermediate Technology Publications, London, UK.

Scoones, I. and Thompson, J. (eds.), 2009. Farmer First revisited. Practical Action Publishing, Rugby, UK.

Sen, A., 1989. Development as capability expansion. Journal of Development Planning, 19, 41-58.

Sen, A., 1999. Development as freedom. Oxford University Press, Oxford, UK.

Shinn, T., 2000. Axes thématiques et marchés de diffusion. La science en France, 1975-1999. Sociologie et Sociétés, 32 (1), 43-69.

Spielman, D., Ekboir, J. and Davis, K., 2009. The art and science of innovation systems inquiry: applications to Sub-Saharan African agriculture. Technology in Society, 13, 399-405.

Wenger, E., 1998. Communities of practice. Learning, meaning and identity. Cambridge University Press, Cambridge, UK.

World Commission on Environment and Development, 1987. Our common future. Oxford University Press, Oxford, UK.

Yung, J.M., and Chauveau, J.P., 1993. Débat introductif. In: Chauveau, J.P. and Yung, J.M. (eds.) Innovation et sociétés. Quelles agricultures? Quelles innovations? Actes du XIVème séminaire d'économie rurale, Montpellier, France, pp. 17-32.

Part I. Thinking innovation differently

Chapter 2. Standards governing agricultural innovation. Where do we come from? Where should we be going?

Lawrence Busch

> *In our investigation we still stress too much the goal of increased productivity as our great task. We still have too much faith in knowledge of the physical and biological facts and principles as all sufficing. There should be searchings of heart as to our policies and programs. Are they adequate to the needs of the new epoch?*
> Kenyon Butterfield (1918: 54)

2.1 Introduction

As noted in the epigraph above, a century ago, Kenyon Butterfield, then President of Massachusetts Agricultural College, asked whether the goal of increased productivity was adequate to the 'new epoch.' But Butterfield's question was largely ignored. For much of the past 300 years, productivity has been the central, and usually unchallenged, goal of agricultural research. Standards and accompanying measures of many kinds were developed to define productivity. Like all standards those for productivity govern our behaviour, shape our institutions and actions, and allow us to measure 'progress.' Yet, today we are faced with Butterfield's question once again. Productivity has not been – and is not likely to be – replaced, but it has been tentatively supplemented by a tangled and sometimes contentious variety of goals often labelled as 'sustainability.'

In this paper what I shall attempt to do is first is to make a few observations about standards. Then, I shall attempt to show how productivity became enshrined in agricultural research as the central goal and how standards and measures were used to describe and indicate progress toward that goal. Next, I will show that the current challenges facing humanity are forcing us to (1) rethink the adequacy of the goal, (2) rethink the standards we use to define and measure it, and (3) to develop new goals, standards, and measures that respond to new challenges. Finally, I will conclude by suggesting how we might begin the difficult task of acting sustainably.

2.2 Standards: building realities

Most of us, whether scientists or laypersons, find standards rather boring. We want to get on with the business at hand. Yet, standards are precisely what permit us to do that, by routinising all sorts of processes, products, and practices (Bingen and Busch, 2005). A breeder who wishes to develop an improved cultivar must have standards (and related measures) in order to know, for example, if the F1 generation constitutes an improvement. A proteomics researcher cannot contribute to our knowledge of plant proteomes without following the standards for such research (Evans, 2000). A nutritionist cannot calculate the nutritive value of a given food without being able to calibrate a variety of instruments using known 'reference materials' (National Institute of Standards and Technology, 2007). And, quite obviously, standard weights and measures are necessary to both good science and global commerce.

But standards do more than provide us with procedures, rules, requirements, specifications, and the like. Standards are ontological devices; they are the recipes by which we create realities (Busch, 2011). Put differently, once a standard has been promulgated and established, if it is of any value at all, it becomes taken-for-granted, natural, the 'obvious' way to do things. It tends to rule out alternatives as irrelevant, beside the point, poor manufacturing, medical, or scientific practice, inadequate. In short, it allows us to create an always partial and incomplete social and natural order, until such time as the standard needs to be revised to take into account new policies, new inventions, new practices, and new discoveries. Thus, for several decades the 'central dogma' told us that genes were what mattered, that any other base pairs were merely 'junk DNA' and of no scientific or social significance. More recently, the central dogma has been replaced with a new standard that permits the development of the so-called 'omic' sciences, i.e. genomics, proteomics, metabolomics, etc.

For much of the last 300 years productivity has been enshrined as the 'gold standard' for agricultural research.[1] The recent and surely well-intentioned statement by FAO Director-General Jacques Diouf that crop productivity has to be increased by 50% by 2050 is paradigmatic of this. The standard has remained largely unchallenged until quite recently, although the means used to measure it have shifted over time. What follows below might be seen as a brief review of productivity standards.

2.3 Before agricultural research: the farmer as experimenter

Agricultural research began nearly as soon as people, whether through population pressure or a desire to remain sedentary, began to practice agriculture. Very quickly, farmers began to select seeds and plants on the basis of the hypertrophy of the edible parts. This activity was certainly not research as we would currently define it, but it was a form of experimentation, of trial and error. It resulted in the creation of most of the crops that we currently plant (Anderson, 1967 [1952]). Indeed, before the invention of scythes, sickles, and other harvesting tools, even grains were harvested individually upon maturity (Haudricourt and Hédrin, 1987). This permitted the careful inspection and sorting of each grain, saving the largest for replanting the following season. To the extent that standards existed, they were highly

[1] Similarly, for much of the last century, gross domestic product (GDP) has been the productivity standard by which national economies have been measured.

localized, seasonally variable depending on the quality of the harvest in a given year, rarely formalized, and typically set by the decisions of individual farmers or perhaps the farmers in a particular village. Sustainability was ensured by following the practices of one's ancestors, by engaging in usually ritualistic imitation of what had been done by the members of the previous generation, and by making fields resemble (what were defined as) natural landscapes (King, 1911). In most instances this worked, although not so much because the gods had viewed the rituals favourably as because the transformations of the landscape wrought by humans were relatively minor, and because responses from the natural world were equally minor.

In a few instances agricultural practices did lead to *long-term* decline rather than to sustainability, but the very long-term nature of this decline meant that it was invisible to those living at a particular time in a particular place (Diamond, 2005). In general terms, there were two major threats to sustainability: declining soil fertility and declining water availability. Importantly, both soil fertility and water availability were always socio-natural phenomena, i.e. they were the product of both human practices and natural processes (Reboul, 1977). The former usually occurred as a limited amount of land was overused to feed a growing population. The latter usually occurred when weather patterns changed – consider the failed early settlements on the American Great Plains – or (more rapidly) when the social organization necessary to maintain irrigation canals collapsed as it did in ancient Mesopotamia. As historian William McNeill (1963, 32) puts it, 'in general, the more elaborate Mesopotamian water engineering became, the heaver became the tasks of maintenance and the greater the chance of sporadic breakdown.'

While farmer trial and error was a slow and arduous process of plant improvement, with numerous setbacks and errors, it permitted the selection and modification of all the plants we cultivate today. Standards for yield were highly localized in character but most emphasized the size of the edible parts. This ultimately led to our dependence on a relatively small number of plant species (as well as their dependence on us), as low-yielding plants were abandoned in favour of those with higher yields. Hence, rice became the basis for the diet in much of Asia, maize was central to the diet in what is now Latin America, and wheat (and to a far lesser extent rye and oats) replaced the millets, spelt, and einkorn in Europe. Much later, as European colonists began to move plants with them around the world, wheat became the favoured crop of large parts of both the North and South American plains. Similarly, in the African wet tropics, maize replaced sorghum as the favoured crop. Without this concern of our ancestors, the more formalized research of the last several centuries would likely not have yielded much.

2.4 Agricultural research as public good, 1600-1980

2.4.1 The rise of botanic gardens

The first wave of what would today be recognizable as agricultural research began with the creation of (modern) botanical gardens. In the Western world the first botanical gardens were created in the 16th century.[2] They had several interrelated purposes:

First, they were to provide a kind of living encyclopaedia of plant knowledge. In the ideal botanic garden, at least one specimen of each species of plant would be grown. Those plants from exotic climes would be grown in glasshouses where temperature and humidity could be controlled. The most famous of these was and remains the Palm House at Kew, where palm trees from throughout the British Empire were grown to maturity. Such living collections greatly enhanced the study of natural history and plant taxonomy, supplementing the botanical codices of earlier centuries.

Second, they would provide for the study of medicinal plants. Indeed, many of the early gardens were of more interest for the medicinal plants they contained than for any agricultural purpose. Indeed, the garden established in France by Louis XIII was known at the *Jardin Royal des Plantes Médicinales* and only expanded beyond the medicinal in 1718 (Bartélemy, 1979). Moreover, at the time it was believed that all plants had medicinal properties. And, in an age in which medicines were as likely to kill as to cure you, at least growing the medicinal plants allowed one to know their origins.

Third, they would allow humankind to re-establish control over nature by recreating the Garden of Eden. Indeed, the Portuguese explorers were convinced for more than a century that the Garden of Eden itself would be found in Brazil (Prest, 1981). Failing that, the attempt was made to recreate the Garden by collecting all species of plants in one location. Here, the Baconian promise of 'reading the Book of Nature' would be fulfilled; God's natural work would be read much as the Bible – God's written work – was read.

Finally, they would allow the 'acclimatization' of plants and their replanting in places other than their place of origin, thereby supporting the aims of the colonial powers. Hence, Britain had its Kew Gardens (ca. 1700), France the *Jardin des Plantes* (1626), the Netherlands the *Hortus Botanicus* (1638), Spain the *Real Jardin Botanico* (1794), and so on. What Calestous Juma (1989: 48) suggested for Britain, was equally true for the other colonial powers: 'On the whole, the British Empire expanded largely as a result of the application of botanical knowledge, technical change and institutional organization to agricultural production'.

[2] Gardens of various sorts have existed since ancient times. What distinguished the new wave of gardens was their link to the beginnings of modern science, itself inextricably linked to the quest for empire and the spread of Christianity.

In short, the botanical gardens took the trial and error approach practiced by farmers for millennia and standardized, systematized, and globalized it. Plant hunting expeditions were sent to the far reaches of the globe to bring back specimens potentially valuable for the colonies. Hence, standards for productivity were centred around the ability of a plant indigenous to colony A to produce an abundant crop in colony B, thereby valorising the latter colony. The result was that the highly lucrative production of (1) stimulant crops, e.g. coffee, tea, cocoa, tobacco, sugar, and (2) industrial crops, e.g. rubber, jute, indigo, was systematically spread to the various European colonies around the world (Brockway, 1979).

2.4.2 From gardens to experiment stations

The Botanic Gardens were quite effective in transferring plants from one continent to another, but their trial and error approach left much to be desired when it came to crop improvement. It was not until the rise of a new kind of institution in the nineteenth century, the agricultural experiment station, that the experimental field trial was standardized and institutionalized. Of particular importance was Justus Liebig's realization that minerals in the soil were directly related to plant growth, and that therefore artificial fertilizers could be produced. With generally enthusiastic support of each of the major world powers of the day, the number of experiment stations exploded from ca. 20 prior to 1851 to nearly 600 by the end of the century and ca. 1500 by 1930 (Busch and Sachs, 1981).

Over time experiment stations introduced a wide range of innovations to the research process. Among them were (1) the comparative study of the performance – as measured in standard units of yield of edible parts – of multiple varieties of the same crop, (2) the use of standard 'control groups' in experimental comparisons, (3) the gradual inclusion of a multitude of (standardized) variables in experimental design, including soil characteristics, plant nutrition, water, latitude, altitude, cultural practices, etc., and (4) the use of standardized statistics to compare results over time and space.

Although there were a few exceptions, in general, like botanic gardens, experiment stations remained creatures of the State.[3] There were several reasons for this: First, most farmers could not afford the rather considerable cost of running an experiment station. Second, even those who could quickly found that there was no way to capture the returns to research; neighbours would soon adopt the innovation without having paid for its development. Third, and probably most important, in the 'metropolitan' countries feeding the growing cities and their working classes was seen as essential to political stability and imperial strength. A matter of such consequence could hardly be left to chance; State control was a necessary condition of social stability.

[3] Exceptions tended to be focused on particular (plantation) crops (e.g. Grammer, 1947).

The same may be said of the International Agricultural Research Centres. They were established in mid-twentieth century as international State projects to spread a green revolution that would counter red revolutions (e.g. Perkins, 1997). They, too, developed products that did not allow the private capture of returns. They were focused almost single-mindedly on increasing productivity. And, they measured their successes in much the same way as national agricultural research organizations – in terms of yield per unit area of edible parts.

2.5 Agricultural research as private good, ca. 1980 – present

Agricultural research began to change markedly in the 1980s. Several critical yet seemingly minor changes took place that markedly shifted the landscape for agricultural research.

2.5.1 Intellectual property (IP) in plants

During the nineteenth century it was assumed worldwide that living things were not patentable, although a small exception was made for certain microorganisms. In 1930, US lawmakers allowed for plant patents; plants reproducible from cuttings – mainly flowers and fruit trees – were given their own IP protection. In the latter part of the century plant variety protection was introduced, a form of IP that permitted plant breeders (in point of fact, private seed companies) to protect new varieties developed through conventional breeding. Already, some observers began to argue that this would transform agricultural research. Said one: 'if international considerations force the introduction of breeders' rights on the United States, it is predictable that government institutes will gradually withdraw from breeding such crops, and confine their activity to basic research' (Fejer, 1966: 4).

Much more recently, as a result of court decisions in the United States, new projects of positive law in the European Union, and the signing of the Trade-Related Aspects of Intellectual Property Rights (TRIPS) agreement (a part of the World Trade Organization), plants and animals became fair game for *industrial* patents. Indeed, since plant variety protection continues to exist, in some sense, property in plants is now better protected than industrial property.

But unlike mechanical or even electronic components, where designing around the patent is commonplace and – at least in principle – there is an infinite, or at least extremely large number of ways of accomplishing the same task, the very nature of living organisms is such that this is difficult or impossible. Moreover, as countless critics have pointed out, ownership of a single gene, such as one conferring resistance to glyphosate (Roundup®), gives the patent owner control of the entire plant. See Box 1 for an alternative approach.

2.5.2 Private investment in plant research

In large part because the intellectual property regime shifted (or was about to shift), companies began to invest heavily in areas of agricultural research previously seen as of little interest. Indeed, nearly every major agrichemical company and quite a few pharmaceutical and oil companies purchased at least one seed company in the expectation that this was a potential replacement for the declining profits in bulk chemicals and that failure to do so was tantamount to being left out of the new opportunities. While the bonanza profits that were promised never materialized, and quite a few oil and pharmaceutical companies later sold off their seed and agricultural divisions (e.g. Novartis sold off its agricultural division as Syngenta), a small number of comparatively large companies did remain engaged in plant research.

Moreover, according to a US Department of Agriculture report: 'private industry spent at least $3.4 billion for food and agricultural research in 1992, compared with $2.9 billion in the public sector. More than 40% of private agricultural R&D is for product development research, compared with less than 7% of public agricultural research' (Fuglie *et al.*, 1996: iii).

Using global data for ca. 2000, Pardey *et al.* (2006) note that the private sector accounts for 36% of all research expenditures worldwide, and 54% of research expenditures in high-income nations.[4] Given that most agricultural research is conducted in high-income nations, the larger figure better illustrates the current dominance of the private sector in agricultural research. That dominance means, among other things, that the private sector – for better or for worse – largely sets the research agenda.

2.5.3 Shifts in the nature of public research

Public research shifted in many critical ways: a shift from block grants to competitive grants, subtle but profound changes in the rewards for scientists, a shift in the skills of scientists, a decline in support for extension activities, an increased flow of private funds into public institutions, and a virtual collapse of public support in some nations. Let us examine each of these in turn.

a. Block grants to institutions were a commonplace phenomena since the inception of agricultural experiment stations in the nineteenth century. A given institution, whether a research agency or higher education institution, received an annual appropriation from the state that was used to further agricultural research. The specific uses for these funds were generally decided upon by the director of the institute, with varying input from scientists employed there and sometimes from farm clientele. This was done for several reasons, not

[4] In France, the United Kingdom, and the United States, the private sector spent 75, 72, and 52 percent of the total, respectively, as of 2000.

Box 2.1. From commodification to commonization of seeds.
Pieter Lemmens and Guido Ruivenkamp

Since the origin of agriculture, seeds have been understood as common goods and were freely exchanged among peasants in all regions of the world. As such, agricultural innovation has always been based on cooperation and sharing. This situation has changed radically during the last two centuries, during which time innovation has been gradually taken over by public institutions, and in the last three decades even more so by private industries. Today, increasingly, the public and private join in order to devote their combined research efforts to the development of varieties attuned to industrialized agriculture embedded in global food chains. Seeds have been transformed from commons to commodities, produced for profit and designed to bring agricultural production under the command of agrifood companies.

The public and, a fortiori, private takeover of agricultural innovation has led, as many have pointed out, to a process of proletarianization among users (farmers and producers) in the sense of an expropriation of their knowledge and knowhow, a loss of their ability to take care and a disengagement of their responsibility for the breeding and cultivation of crops. These have been delegated now to scientists, whose research agenda is increasingly submitted to the dictates of corporate capital. Capital, however, is by its very nature incapable of care and responsibility since it is principally interested in profit-making and structurally indifferent to social and ecological values.

To restore care and responsibility in agriculture and to make it more sustainable and equitable, it is necessary not so much to revive the public sector (increasingly dependent on the logics of the private sector) but to revive the commons, i.e. to find ways to re-commonize innovation and counter the processes of proletarianization and commodification of seeds and germplasm with processes of commonization. Open source initiatives like CAMBIA-BiOS, which proffer a commons-based approach to agricultural innovation, are important steps towards this goal and may be increasingly embraced and elaborated by peasants, civil-society organisations, farmers and scientists.

The guiding principle of open source based innovation consists in the openness and free availability of the tools of innovation. Its primary orientation is not the creation of exchange value, i.e. profit, but use value, i.e. the permanent expansion of a shared pool of resources. CAMBIA firmly believes that innovation is most efficient and productive not when tools and information are privatized and turned into exclusive property but when they are commonized and made inclusive. Its mission is to democratize innovation and therefore it has developed a suite of tools – like its Transbacter vector technology as an alternative to the patented Agrobacterium vector – and research practices made available under an open source license inspired by the GPL Free Software license orginally created by Richard Stallman for the software industry. Through this license, CAMBIA aims to create a protected commons of technologies and capabilities that cannot be appropriated by private entities. As such, it explicitly engages in the (re-)empowerment of users, not owners (Lemmens, 2010; Ruivenkamp and Jongerden, 2010).

least of which was that experiments required a rather significant amount of infrastructure. For example, experiments on cattle or other farm animals required that a herd be maintained for an indefinite period of time; similarly, plant experiments required that large areas be maintained for field experiments. Moreover, management was, and remains, quite complex, with considerable effort involved to ensure that the treatments from one experiment do not influence the following one on the same field. However, in recent years this form of funding has fallen out of favour. This is not least the result of the neoliberal obsession – an obsession that began to hold sway in much of the world – with introducing competition into all activities (Bourdieu, 1998). Given that most competitive grants rarely last more than 3 to 5 years, long-term research has suffered. Moreover, at least one recent study suggests that this shift has actually lowered crop productivity gains (Huffman and Evenson, 2006).

b. Furthermore, while a century ago the results of nearly all agricultural research were published in bulletins – usually in-house publications of varying length – by the second half of the 20th century most research results found their way to scientific journals (Busch and Lacy, 1983).[5] This was doubtless partially the result of growing professionalism of scientists, as well as a realization of the large spillover effects in agricultural research.[6] However, in more recent years the reward system of public research institutions shifted so as to emphasize the production of large numbers of scientific journal articles (often as opposed to the generation of products such as plant varieties, or practices such as means for better managing planted fields or farm animals), and in some institutions to citations to those articles. This has meant that conventional plant and animal breeding, as well as any other scientific activity that does not yield short-term results has been downgraded if not abandoned. Indeed, in institutions where citation analyses has gained favour, pressure has increased on researchers to publish in the most highly cited journals as well as to submit book manuscripts only to the most well-known presses. This shift in the reward structure offers major advantages to research administrators. In particular, it appears to offer a seemingly objective means of comparing the results of multiple researchers working in different fields of science on topics of varying importance. It appears to offer a measure of the success of a given scientist in the 'marketplace of ideas'. However, there are many serious flaws in such a perspective. Counting articles is easy to do, but in many disciplines it is quite possible for scientists to divide their work into ever finer pieces thereby artificially increasing the number of articles produced. Moreover, the rate of production of articles depends heavily on the type of research conducted. Some experiments can be completed in thirty days, while others take several years. Similar problems apply to the use of citations. First, citation databases are neither complete nor random; they only include certain journals – often excluding those not in English or in highly specialized and/or new fields of study. Second, citation rates are heavily

[5] Already, by the early twentieth century US experiment station directors were concerned that bulletins were inappropriate locations for the results of scientific research. See, e.g. Pearl (1915).

[6] Put simply, results from one location are nearly always applicable elsewhere; they spill over into other geographic spaces that have similar environmental features.

dependent on the number of scientists working in a given field; scientists working in small fields – regardless of the importance of their research – cannot receive as many citations as those working in larger fields. Third, certain papers may be cited for their flaws rather than for their relevance; indeed, the paper on stem cells published in *Science* by Hwang Woo-suk, later determined to be fraudulent, has been cited 198 times according to the *Science Citation Index*. Fourth, there is at least some evidence of journal editors encouraging authors to add citations to their papers so as to increase the citation rate (and presumably the status) for a particular journal. Finally, the value of research may not be recognized by other scientists until some time after it was conducted; thus annual reviews of citations may be misleading at best. But what is perhaps most perverse about the use of citations in this manner is that it actually undermines the scientific enterprise. It encourages the best scientists to work in large fields on well-known, geographically widespread topics that are often only remotely connected to problems faced by practitioners (e.g. Hanafi, 2010). Moreover, high-risk topics are to be avoided. After all, high-risk research is by definition *more likely to fail* and thus to either be unpublishable or only publishable in an obscure journal.[7]

c. The combination of changes in the reward structure and the rise of the burgeoning new field of plant molecular biology led research managers to shift hiring practices. Whereas in the past agricultural research institutes each had a few taxonomists and large numbers of plant breeders, these fields were now seen as passé. When the eminent taxonomist Jack Harlan, who had warned as early as the 1930s of declining genetic resources (Harlan and Martini, 1936), retired from the University of Illinois, he was not replaced. And, by the 1990s there was already a discernable shortage of plant breeders (Frey, 1996). In contrast, molecular biologists – regardless of the quality of their work – have nothing to offer directly to farmers.

d. Moreover, extension activities were allowed to decline precipitously, especially in poor nations but even in wealthy nations such as the United States. To administrators and politicians this made sense since public sector research was generating fewer products that could be used directly by farmers; most products now required 'development' by private companies. Furthermore, the private sector has increased its 'extension' efforts as potential product sales have increased. Moreover, with far fewer farmers in wealthy nations, the justification for public extension activities has been weakened (Wolf and Zilberman, 2001). Indeed, in New Zealand the entire extension service was privatized on a fee for service basis.

[7] Instructive here is the experience of Nobel Prize-winning economist, George A. Akerlof. His now famous article, 'The market for lemons' (Akerlof, 1970), was rejected by the top journals in the field in part because it challenged the status quo (Cassidy, 2009). He might well have been discouraged from publishing it at all in the current reward structure.

e. The discomfort once felt with respect to accepting private funds for research or otherwise getting involved in common projects evaporated;[8] universities and government research laboratories began to see private funds as a new revenue stream, one that would replace 'inadequate' or declining public funds for research. Indeed, private investment in public research institutions increased from a small trickle to a veritable flood. In the United States by 1994 some 20% of research conducted at State Agricultural Experiment Stations was financed by the private sector, even as public expenditures by the federal and state governments stagnated or declined. The now somewhat infamous agreement between the University of California Berkeley and the Novartis Corporation provided a paradigm case for this type of relationship (Rudy *et al.*, 2007).

f. Finally, in some nations public agricultural research has been largely or entirely abandoned. This is the case for the United Kingdom, where the prestigious Cambridge Plant Breeding Institute was sold off to the private sector, and numerous other agricultural research institutes were closed. In numerous other nations, public research has either stagnated or been largely abandoned (see, e.g. Echeverria, 1998).

2.6 Supermarkets, processors, and the standards revolution

But the story does not end here. For the last decade has seen an astonishing and largely unpredicted growth in the scope and power of downstream actors in food supply chains. In particular, a small number of well-organized, large-scale supermarket chains has penetrated markets previously considered too small, too poor, or too disorganized to be of interest (Dries *et al.*, 2004; Reardon *et al.*, 2003; Weatherspoon and Reardon, 2003). This transformation has brought downstream issues into focus – including nutrition, food quality, and sustainability – but it also has a number of serious pitfalls associated with it. In particular, in their zeal (1) to protect and bound off their supply chains (Busch, 2007), and (2) to avoid the reputational risks involved in selling contaminated food, supermarkets have begun to impose a plethora of standards on their suppliers (Fulponi, 2006; Mutersbaugh, 2005). Farms and processors must be efficient and productive, but more and more they are being asked to be efficient and productive in certain highly specified ways. Hence, a wide array of certifications and audits have been mandated by retailers. While in principle these standards, certifications, and audits are voluntary, in practice they are mandatory – violated on penalty of failure to sell one's crop.

What these myriad standards do is to freeze innovation on the farm. Collectively, and doubtless with no malice on the part of the vast majority of supermarket chains, they impose heavy burdens on suppliers. Suppliers must use particular cultivars, meet strict cosmetic quality requirements, employ particular field practices, fertilizer and pesticide spray

[8] A 1928 US report argued strongly that commercial funding should only be accepted when the research was of general public importance and then only when the funds were given to the institution rather than to a particular researcher (Barre *et al.*,1928).

regimes, harvest and deliver on defined dates, and pack products in boxes of defined sizes. Together, these standards have the (largely unintended, but nevertheless problematic) effect of channelling research in certain directions. Since they are not integrated into a vision of a more sustainable food production and distribution regime, they may actually hamper the achievement of enhanced sustainability.

In sum, as a consequence of (1) a declining public share, and in some cases unstable or declining funding in absolute terms, (2) changes in funding mechanisms from block grants to competitive grants, (3) changes in the reward system that discourage scientists from research that leads directly to practical application, (4) declining support for public extension activities, and (5) the rise of retailer-led supply chains, the private sector gradually but surely has become the central guiding element in agricultural research. Put differently, research has become largely a private good. Even public institutions have tended to reward that research that leads to new revenue streams – whether through patents or spin-off enterprises – rather than research that serves the public at large. Moreover, the combination of stronger intellectual property rights and related regulation with private sector dominance are reducing the classic spillover effects of agricultural research. Put differently, in the past the products of research were often easily adapted to other locales, while today that is far less likely. Pardey *et al.* (2006: 18) explain: 'notably, rich countries are reorienting their agricultural R&D away from the types of technologies that are most easily adapted and adopted by developing countries. In addition, intellectual property rights and other regulatory policies – including biosafety protocols, trading regimes, and specific regulatory restrictions on the movement of genetic material – are increasingly influencing the extent to which such spillovers are feasible or economic'.

2.7 The coming storm

What I have just described would be of little consequence were it not for the fact that the very standards and measures in use to define and calculate productivity gains are actually leading us astray. Consider the issues currently facing both the scientific community as well as everyone on earth:

- About one-fifth of the world's population remains undernourished despite considerable increases in productivity.
- We are more reliant than ever on non-renewable fossil energy to maintain productivity levels – not merely on the farm, but in the production of inputs, and the distribution of outputs.
- The limiting macronutrient in plant nutrition is phosphate. Phosphate supplies are quite limited, yet phosphate use efficiency is quite low.
- Obesity is a significant and growing problem, not just in industrial nations but even in middle-income and poor nations.
- The current agrifood regime is responsible for a wide range of environmental problems including nitrate runoff from farms, soil erosion and depletion, and a reduction in biodiversity.

- Climate change is likely to shift the conditions of production in many regions, leading to a mismatch of crops and water needs.
- The terms of trade between industrial and poor nations continue to decline.

Each of these issues alone is clearly a barrier to developing a sustainable food supply. But equally important, *current standards and measures for productivity tend to obscure these disconcerting issues.* They perpetuate two myths: (1) that we can resolve the problems of food and agriculture solely by increasing productivity, and (2) that there is only one path to sustainable development that everyone and every nation must follow. These are certainly myths, but they are real myths. That is to say, they are real because, as W.I. Thomas (1928: 572) suggested, they are 'real in their consequences'. At best, they send the wrong signals to scientists and others in food chains, at worst they mask grave vulnerabilities and barriers to moving toward a more sustainable means of feeding ourselves (Tansey and Rajotte, 2008; Weis, 2007).

To appreciate the problems facing us, consider a simple thought experiment. Let us assume that, through some extraordinary feat of scientific achievement, we were able to double the productivity of all major crops beginning next year. In other words, using the same quantity of inputs we were able to produce twice the quantity of food. This would *not* result in a halving of the cost of food to consumers since only ~20-25% of the cost of food is attributable to on-farm activities. Hence, food prices would decline by only ~10-15%. This would clearly be a significant improvement for those at the bottom of the income ladder, but the vast majority of hungry people would still remain hungry. Moreover, all of the other problems associated with the food production arrangements would remain. We would still have to deal with declining resources, environmental pollution, obesity, etc.

2.8 Widening the scope of agricultural research

What all of this suggests is that there is a critical need to widen the scope of agricultural research. See Boxes 2 and 3 for examples of alternative approaches. Productivity will have to become one among several goals. Standards and measures will need to be developed to grapple with the wicked problem (Batie, 2008; Rittel, 1972; Rittel and Webber, 1973) of sustainability. Sustainability is a wicked problem precisely because (1) there is no optimal solution but rather tradeoffs among various objectives, (2) not all parameters are either known or equally subject to specification, (3) conflicts are bound to occur as a great deal is at stake and there will be both winners and losers, (4) any given aspect of the problem is itself another problem, and (5) no demonstrably correct 'solution', but only iterative 'improvements' are possible.

Consider that within the agricultural research community in recent years we have had calls for research on problems of global warming, rural development, environmental improvement, economic growth, sustainability, and even public health (Busch, 2009). Whereas in the past we could bracket these issues – consider the initially hostile reaction of the entomological

Box 2.2. Sustaining standards in a complex world: the case of SRI.
Shambu Prasad Chebrolu

The first decade of the 21st century has witnessed an unprecedented agricultural distress resulting in over 200,000 Indian farmer suicides, most of them from areas that were hitherto part of the green revolution states. This calls for a rethink on the paradigm of agricultural science that has privileged productivity enhancement over poverty reduction or ecological sustainability. Surprisingly though, the same period has seen a steady spread of agroecological practices and innovations such as SRI or the system of rice intensification. SRI, is a set of ideas and insights that originated in Madagascar through the systematic experimentation by a French priest who put together a set of six principles for rice cultivation (wider spacing of rice plants, transplanting young and single seedlings, maintaining moist but unflooded rice fields, soil aeration through mechanical weeding, and use of soil organic matter) that together creates a better growing environment. SRI improves yields, enhances soil health, and reduces the need for inputs (seeds, water, labour) without changes or improvements in variety.

SRI though has had a cold, if not hostile, response from the rice research establishment. A fixation on on-farm productivity and a controversy over claims of yields exceeding the biological maximum of the rice plant has meant that SRI has not been seen as one of the alternatives for rice production. Proponents of SRI have pointed to wider impact of SRI that, while including higher crop yields and productivity, has also focused on enhanced farm incomes, especially of poor farmers, the resource conserving aspect of the innovation such as reduced water requirement, greater pest and disease resistance and resilience to climatic stresses such as drought, storm damage and extremes of temperature. Global estimates on the number of farmers who are practicing SRI is between 2 and 3 million. Controversies over the yield potential and the inability of replicating and assessing the synergistic effect in laboratory conditions, often on soils that are devoid of microbial activity, has meant that scientists have paid little attention to farmer experimentation that has resulted in the rapid spread of SRI on farmers fields. Public investment in SRI research and extension has been low with little private investments. SRI however offers an alternate architecture of knowledge being placed in the public domain and as open source enabling the new commons in agriculture.

The few researchers who have worked on SRI have pointed to the possibility of the physiology of the rice plant being different under SRI and the need for newer standards to assess SRI. The spread of SRI has largely been due to the internet, providing an opportunity for a new kind of public research with free exchange of information and ideas among researchers, farmers, donors and civil society organisations. New electronic groups like the popular SRI India Google Group with over 400 members and newer groupings such as learning alliances of multi-actor coalitions have driven the spread of SRI. SRI principles have also been extended to other crops. The happenings in the field point to the need for a closer look at standards such as productivity and yields as the sole defining criteria for agricultural growth and the need to push for a pluralistic definition that includes farm incomes, soil health and resource conservation strategies (Chebrolu and Sen, 2010).

Box 2.3. An example of reorientation of agricultural research to benefit small producers in Brazil.

Eliane de Carvalho Noya, Bernard Roux and Geraldo Majella Bezerra Lopes

Encouraged by the public policies of the Lula government, the agricultural research institute for Pernambuco state (IPA), in north-eastern Brazil, has recently implemented a method of agricultural development based on the concept of a collective learning and research unit (UPAC). It consists of integrating, in close collaboration with organizations of small family farmers, the knowledge and experience of all local actors concerned with food production: the farmer, his family, the extension worker, the scientist, the teacher and the local leader. To provide an alternative to the old ways of 'disseminating technical information' and facilitate this integration, collective activities of multidisciplinary research and learning are undertaken through site visits, participatory observations, interviews with key individuals, discussion sessions for diagnosing problems and the monitoring of the real situations of farming communities. The premise being that this knowledge will be useful in developing new products, processes or services or in improving existing ones. Endogenous and exogenous technological innovations are identified, systematized, improved and validated on themes as diverse as aquaculture, viticulture, horticulture, polyculture, economic solidarity, associativism, etc. This permits the identification of geographical, socio-economic, cultural and other elements which are usually overlooked by highly specialized researchers.

The UPAC is based on a systemic approach consisting of the following objectives:

- the participation of all in building knowledge about food production in view of an overall goal of food security for the community;
- social justice and equal employment and income opportunities, irrespective of sex or social group;
- sustainability of the development process for the benefit of present and future generations;
- conservation and appreciation of the value of local resources;
- individuals, in their role as citizens, becoming masters of their own destinies.

Results were found to be encouraging. The mutual learning process between the extension workers, the scientists and the farmers was particularly effective. The latter found ways of resolving their problems through interaction with the other actors, local authorities and funding agencies. Scientists and extension workers radically altered their behaviour. In this way, all UPAC participants helped advance research, create knowledge and information and develop technologies by being part of a joint endeavour for agricultural and local development, all thanks to the improvement in human and social capital that this approach engenders. A question, finally, of changing the approach to research and its methods and contents (De Carvalho *et al.*, 2010).

community in the 1960s to suggestions that chemical control of agricultural pests might be problematic – this is a luxury that we can no longer (and perhaps never could) afford.

Rethinking and more importantly re-enacting agricultural research will doubtless be upsetting for some in the scientific community. I speak specifically of those scientists who follow Kaplan's (1964) law: Give a boy a hammer and he finds that everything needs hammering. As things currently stand, most agricultural scientists spend five years or more learning all about one narrow aspect of agricultural research; many mistakenly believe that nearly all problems can be solved by that discipline – whether it is plant breeding, molecular biology, or agricultural economics. We need to revamp agricultural higher education so that even as students learn the details of one discipline, they are conversant in others.

We need to begin to do research with value chains and foodsheds in mind rather than single-mindedly focusing on productivity. To examine what that might mean, consider the case of phosphorus. The world is in very short supply of phosphorus, and it is essential for plant nutrition. Yet, we waste vast quantities of it every season (e.g. Quinton *et al.*, 2010). A truly interdisciplinary research program might ask:

- How can phosphorus use efficiency by plants be improved?
- What policies might be put into place that would encourage phosphorus conservation and recycling?
- What changes in fertilizer production and application might be made that would conserve phosphorus?
- What costs/impacts are there to phosphorus pollution to 'downstream' actors?
- Might phosphorus use in other applications be reduced so as to leave more for agricultural uses?

Yet, even this would remain a largely 'upstream' problem in that it would not grapple with the fundamental problem of sustainability. A program of research of this sort would have to confront the entire food value chain from input production to final consumption. Moreover, since there is already sufficient food produced to feed everyone on the planet, it would logically put more initial emphasis on post-harvest issues than on increasing productivity. It would have to take seriously the 'right to food' as enshrined in international law (United Nations, 1948: §25 (1)). Such a research program would ask:

- How might the very significant global post-harvest food losses be reduced, by changes in transport, processing, marketing and other practices, and by policies designed to reduce such losses?
- How might the differing food needs of small farmers, farmworkers, and the urban poor be addressed? Specifically, what changes in food chain practices, technical changes, and policy shifts might ensure a safe, sustainable, sufficient, nutritionally adequate food supply?
- What physical, legal, policy, and other infrastructural needs must be met to ensure a sustainable food supply for all?

Addressing these wicked problems will require political courage and leadership as well as more than a little concerted effort that transgresses the real but artificial boundaries of the disciplines. The scientific community will need to develop new measures, new standards, to address this multi-faceted challenge. The increasingly quixotic quest for a singular, universal solution to the problems of food and agriculture will need to be abandoned. Public sector science will need to regain control of the research agenda and redirect it in ways that promote food security. These are not easy tasks to accomplish. There will doubtless be failures along the way. But for those reasons alone, we cannot wait. To paraphrase the American poet, Robert Frost, we have 'miles to go before we sleep'.

References

Akerlof, G.A., 1970. The market for 'lemons': quality uncertainty and the market mechanism. Quarterly Journal of Economics, 84, 488-500.

Anderson, E., 1967 [1952]. Plants, man and life. University of California Press, Berkeley, CA, USA.

Barre, H.W., Call, L.E., Kendall, J.C., Mooers, C.A., Allen, E.W., and Jardine, J.T., 1928. Report of the committee on experiment station organization and policy: continuity in research. In: Proceedings of the 42nd Annual Convention of Land-Grant Colleges and Universities, Washington, DC, USA, pp. 203-205.

Bartélemy, G., 1979. Les jardiniers du roy: petite histoire du jardin des plantes de Paris. Le Pélican, Paris, France.

Batie, S.S., 2008. Wicked problems and applied economics. American Journal of Agricultural Economics, 90, 1176-1191.

Bingen, J. and Busch, L. (eds.), 2005. Agricultural standards: the shape of the global food and fiber system. Springer, Dordrecht, the Netherlands.

Bourdieu, P., 1998. The essence of neoliberalism. Le Monde Diplomatique (English Edition), Available at: http://mondediplo.com/1998/12/08bourdieu.

Brockway, L.H., 1979. Science and colonial expansion: the role of the British Royal Botanic Gardens. Academic Press, New York, NY, USA.

Busch, L., 2007. Performing the economy, performing science: from neoclassical to supply chain models in the agrifood sector. Economy and Society, 36, 439-468.

Busch, L., 2009. What kind of agriculture? What might science deliver? Natures, Sciences, Sociétés, 17, 241-247.

Busch, L., 2011. Standards: recipes for reality. MIT Press, Cambridge, MA, USA.

Busch, L. and Lacy, W.B., 1983. Science, agriculture, and the politics of research. Westview Press, Boulder, CO, USA.

Busch, L. and Sachs, C., 1981. The agricultural sciences and the modern world system. In: Busch, L. (ed.) Science and Agricultural Development. Allanheld, Osmun, Totawa, NJ, USA, pp. 131-156.

Butterfield, K.L., 1918 [1917]. The Morrill Act institutions and the new epoch. In: Proceedings 31st Convention Association of American Agricultural Colleges and Experiment Stations. Washington, DC, USA, pp. 43-59.

Cassidy, J., 2009. How markets fail. the logic of economic calamities. Farrar, Strauss and Giroux, New York, NY, USA.

Chebrolu, S.P. and Sen, D., 2010. The new commons in agriculture: lessons from the margins and SRI in India. In: Innovation and Sustainable Development in Agriculture and Food – ISDA 2010, Montpellier, France. Available at: http://hal.archives-ouvertes.fr/hal-00521398/fr/.

De Carvalho Noya, E., Roux, B. and Majella Bezerra Lopes, G., 2010. Changer est nécessaire, changer est-il possible? interaction entre acteurs et construction de nouvelles pratiques en science et technologie pour l´agriculture familiale dans l'Etat de Pernambouc, Brésil. In: Innovation and Sustainable Development in Agriculture and Food – ISDA 2010, Montpellier, France. Available at: http://hal.archives-ouvertes.fr/hal-00521362/fr/.

Diamond, J., 2005. Collapse: how societies choose to fail or succeed. Viking, New York, NY, USA.

Dries, L., Reardon, T., and Swinnen, J.F.M., 2004. The rapid rise of supermarkets in Central and Eastern Europe: implications for the agrifood sector and rural development. Development Policy Review, 22, 525-556.

Echeverria, R.G., 1998. Agricultural research policy issues in Latin America: an overview. World Development, 26, 1103-1111.

Evans, G.A, 2000. Designer science and the 'omic' revolution. Nature Biotechnology, 18, 127.

Fejer, S.O., 1966. The problem of plant breeder's rights. Agricultural Science Review 3rd quarter, 1-7.

Frey, K., 1996. National plant breeding study – I: human and financial resources devoted to plant breeding research and development in the United States in 1994. Iowa Agriculture and Home Economics Experiment Station, Ames, IA, USA.

Fuglie, K., Ballenger, N., Day, K., Klotz, C., Ollinger, M., Reilly, J., Vasavada, U. and Yee, J., 1996. Agricultural research and development: public and private investments under alternative markets and institutions. USDA, Washington, DC, USA.

Fulponi, L., 2006. Private voluntary standards in the food system: the perspective of major food retailers. Food Policy, 31, 1-13.

Grammer, A.R., 1947. A History of the experiment station of the Hawaiian sugar planters' association: 1895-1945. The Hawaiian Planters' Record, 51, 177-228.

Hanafi, S., 2011. University systems in the Arab East: publish globally and perish locally vs. publish locally and perish globally. Current Sociology, 59, 291-309.

Harlan, H.V. and Martini M.L., 1936. Problems and results of barley breeding. United States Department of Agriculture, Washington, DC, USA.

Haudricourt, A.G. and Hédrin, L., 1987. L'homme et les plantes cultivées. Editions A.-M. Métailié, Paris, France.

Huffman, W. and Evenson, R., 2006. Do formula or competitive grant funds have greater impacts on state agricultural productivity? American Journal of Agricultural Economics, 88, 783-798.

Juma, C., 1989. The gene hunters: biotechnology and the scramble for seeds. Princeton University Press, Princeton, NJ, USA.

Kaplan, A., 1964. The conduct of inquiry. Chandler, Scranton, PA, USA.

King, F.H., 1911. Farmers of forty centuries. Rodale Press, Inc., Emmaus, PA, USA.

Lemmens P., 2010. Deproletarianizing agriculture – recovering agriculture from agribusiness and the need for a commons-based, open source agriculture. In: Innovation and Sustainable Development in Agriculture and Food – ISDA 2010, Montpellier, France. Available at: http://hal.archives-ouvertes.fr/hal-00539829/fr/.

McNeill, W.H., 1963. The rise of the west. University of Chicago Press, Chicago, IL, USA.

Mutersbaugh, T., 2005. Fighting standards with standards: harmonization, rents, and social accountability in certified agrofood networks. Environment and Planning A, 37, 2033-2051.

National Institute of Standards and Technology, 2007. Standard Reference Materials® Catalog. US Government Printing Office, Washington, DC, USA.

Pardey, P.G., Beintema, N., Dehmer, S. and Wood, S., 2006. Agricultural research: a growing global divide? International Food Policy Research Institute, Washington, DC, USA.

Pearl, R., 1915. The publication of the results of investigations made in experiment stations in technical scientific journals, including the journal of agricultural research. In: Proceedings of the 29th Convention of the Association of American Agricultural Colleges, Berkeley, CA, USA.

Perkins, J.H., 1997. Geopolitics and the green revolution. Oxford University Press, New York, NY,USA.

Prest, J., 1981. The garden of eden: the botanic garden and the recreation of paradise. Yale University Press, New Haven, CT, USA.

Quinton, J.N., Govers, G., Van Oost, K. and Bardgett, R.D., 2010. The impact of agricultural soil erosion on biogeochemical cycling. Nature Geoscience, 18, 311-314.

Reardon, T., Timmer, C.P., Barrett, C.B. and Berdegue, J., 2003. The rise of supermarkets in Africa, Asia, and Latin America. American Journal of Agricultural Economics, 85, 1140-1146.

Reboul, C., 1977. Determinants sociaux de la fertilité des sols. Actes de la Recherche en Sciences Sociales, 17/18, 85-112.

Rittel, H.W.J., 1972. On the planning crisis: systems analysis of the 'first and second generations'. Bedriftsøkonomen, 8, 390-396.

Rittel, H.W.J. and Webber, M.M., 1973. Dilemmas in a general theory of planning. Policy Sciences, 4, 155-169.

Rudy, A.P., Coppin, D., Konefal, J., Shaw, B.T., Eyck, T.T., Harris, C. and Busch, L., 2007. Universities in the age of corporate science: the UC Berkeley-Novartis controversy. Temple University Press, Philadelphia, PA, USA.

Ruivenkamp, G. and Jongerden, I., 2010. Open source and commons in development. In: Innovation and Sustainable Development in Agriculture and Food – ISDA 2010, Montpellier, France. Available at: http://hal.archives-ouvertes.fr/hal-00521937/fr/.

Tansey, G. and Rajotte, T. (eds.), 2008. The future control of food. Earthscan, London, UK.

Thomas, W.I., 1928. The child in america: behavior problems and programs. Alfred A. Knopf, New York, NY, USA.

United Nations, 1948. Universal declaration of human rights. United Nations, New York, NY, USA.

Weatherspoon, D.D. and Reardon, T., 2003. The rise of supermarkets in Africa: implications for agrifood systems and the rural poor. Development Policy Review, 21, 1-17.

Weis, T., 2007. The global food economy: the battle for the future of food and farming. Zed Books, London, UK.

Wolf, S. and Zilberman, D. (eds.), 2001. knowledge generation and technical change: institutional innovation in agriculture. Kluwer Academic Publishers, Boston, MA, USA.

Chapter 3. From concept to emerging practice: what does an innovation system perspective bring to agricultural and rural development?

Bernard Triomphe and Riikka Rajalahti

3.1 Introduction

In the midst of deep, complex environmental and socio-economic change occurring from the local to the global scale, farmers, agribusinesses, and societies must innovate continuously if they are to cope, compete, and thrive. Sustainable agricultural development in particular demands – and depends on – an array of technical, social and institutional innovations.

How best to innovate has been the focus of much debate. For decades, many actors of the agricultural sector have advocated for and invested heavily in the development and strengthening of public research, extension, education, and in improving the links with one another as a way to produce and disseminate the knowledge and array of technologies perceived as necessary for agricultural development. The underlying model for innovation was mostly linear: in such a model, scientists produce and use their knowledge to create new technologies which are then disseminated to the potentially adopting farmers by extension officers. To foster adoption, the latter provide farmers with support in the form of advice, credit, training, etc. (Rogers, 1983). These investments and approaches have had substantial returns, with many farmers and consumers benefiting from them, as evidenced by the much debated yet undisputable impact of the Green Revolution (Evenson and Gollin, 2003). However, they have not always spurred sufficient impact, especially for the poorer, more marginalized stakeholders and environments. Nor have they kept pace with the myriads of demands, challenges and opportunities emerging from an ever wider diversity of actors in a constantly changing environment (Rajalahti *et al.*, 2005, 2008; World Bank 2012).

The limited success of conventional linear approaches to innovation, coupled with the accumulated evidence and learning about how innovation actually takes place, led to the emergence, and more importantly, the gradual endorsement and implementation of approaches related to the innovation systems (IS) concept (Freeman, 1988; Lundvall, 1992; OECD, 1997). Initially applied to technology development in developed economies, the IS concept was eventually applied and adapted to the agricultural sector of both developed and developing countries (World Bank, 2006).The resulting Agricultural Innovation Systems (AIS) perspective has proven very useful, e.g. in assessing the strengths and weaknesses of given innovation processes, systems and environments, and also, increasingly, in planning and implementing actions with the potential to contribute to more functional, dynamic innovation systems and better impact of innovation investments.

This chapter introduces the AIS perspective and highlights its contribution to the analysis and design of agricultural innovation. Firstly, it provides a brief overview of the emergence of the IS concept and perspective, including its links with concepts such as actor-network theory. Secondly, it outlines a series of approaches and instruments which all contribute to operationalizing the AIS perspective and giving it a practical content. Finally, it outlines a number of current challenges related to the use of an IS perspective, while the conclusion mentions a few directions in which future efforts into the AIS perspective might be directed.

3.2 What is an innovation system and where does the IS concept come from?

3.2.1 Brief history of the innovation system concept

Simply put, innovation is the process by which individuals or organizations master and implement the design and production of goods and services that are new to them. It can also be defined as an invention having found a market (Schumpeter, 1942), a useful distinction because it differentiates the mostly 'knowledge' content of innovation (coming up with a new idea) from its application.

When looking back at many innovation histories or trajectories, a recurrent understanding and lesson gained was that 'successful' innovation could often be attributed to the existence of effective networks through which scientists and entrepreneurs from the public and private sectors interacted, learned from one another, shared resources and responded quickly to changing economic and technical conditions (World Bank, 2006). In particular, the success of economies such as Japan in the 1970s and 1980s drew much attention, as other developed countries were keen to emulate Japan. Studying these economies, Freeman (1988) coined the term 'national system of innovation' to refer to the State institutions involved in defining and performing research and innovation policy. In further studies of national innovation systems in the early 1990s, Nelson (1993) and Lundvall (1992) widened Freeman's definition to include actors from the industrial sector and elements of the national context within which research and innovation take place. Throughout the 1990s, the Organization for Economic Cooperation and Development (OECD) increasingly referred to 'framework conditions' in describing the scope of IS: contextual factors such as tax regimes, regulations, laws, culture, and behaviours, broadening the definition of innovation systems still further (OECD, 1997).

Authors such as Latour (1987), Akrich *et al.* (1988), Callon *et al.* (2001) have, for their part, made several key contributions, albeit not always acknowledged, to the present-day IS concept through their seminal work on Actor-Network Theory (ANT). ANT focuses on understanding the emergence, functioning and structure of innovation networks linking a diverse set of actors and objects with which they interact in the process of innovation development. In this view, the innovation process is seen as a result of co-evolutionary and organic reciprocal adjustments between the object of innovation and the surrounding society, which follows a non-linear, unpredictable trajectory. ANT also analyzes the diverse incentives and motivation for innovation, or on the contrary, the obstacles and varied forms of resistance innovation may face over time. ANT furthermore points out to the central role played by mediators, actors with the necessary skills and power to negotiate meaning and common understanding as part of a translation process which is an integral part of the innovation process.

While the IS concept was initially applied at the national scale, it has since been applied at other scales as well: international, regional, local or sectoral. For each scale, the set of actors involved in innovation and the actual context for the system varies. Time is another key dimension in the analysis as innovation systems evolve dynamically over usually rather long periods, as the ANT authors have amply demonstrated (e.g. Alter, 2000). Some IS scholars propose to deal with the time dimension by distinguishing between nascent, emerging, landscape and regime phases (Knickles *et al.*, 2008), or between nascent, emergent and stagnation phases (World Bank, 2006). Besides scale and time, another useful distinction relates to whether innovation arises out of a planned or orchestrated process, or in a more spontaneous, non-orchestrated manner (World Bank, 2006), which more recently has also been referred to by some authors as 'innovation ecosystems' (Fukuda and Watanabe, 2008).

With this historical background, an *innovation system* can be defined as 'a network of organizations, enterprises, and individuals focused on bringing new products, new processes, and new forms of organization into economic use, together with the institutions and policies that affect their behaviour and performance' (World Bank 2006).

While several authors have defined innovation systems slightly differently by focusing on specific aspects (e.g. Freeman, 1987; Lundvall, 1992, or Metcalfe, 1995), all existing definitions have in common to propose a systemic framework for viewing the rather complex web of actors and processes related to technology development (OECD, 1997). In doing so, they recognize two key aspects: the multiplicity of actors involved and their interactions, and the environment in which these actors operate.

3.2.2 The innovation system concept applied to the agricultural sector

The agricultural sector is but one example of how the IS concept can be applied to gain new understanding of how innovation can develop and be nurtured. In an Agricultural Innovation System (Figure 3.1), innovation usually arises through dynamic, open interaction among the multitude of actors involved in growing, processing, packaging, distributing, and consuming or otherwise using agricultural products, allowing to draw on the most appropriate available knowledge. Contributions by research, extension and education are

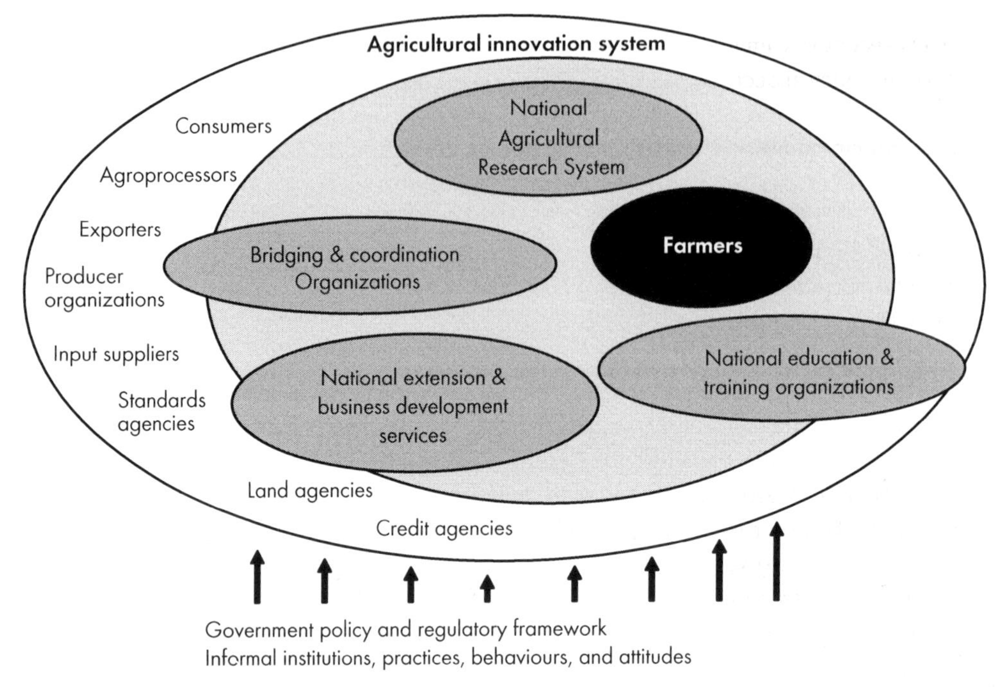

Figure 3.1. A stylized agricultural innovation system (World Bank, 2012).

usually necessary but not sufficient to bring the needed knowledge, technologies, and services to farmers and entrepreneurs. Besides strong capacity in R&D, the ability to innovate is often related to collective action, coordination, the exchange of knowledge among diverse actors, the incentives and resources available to form partnerships and develop businesses, and the existence of conditions that make it possible for farmers or entrepreneurs to use the innovations (Hall *et al.*, 2003; World Bank, 2012).

Compared to earlier approaches and paradigms, the AIS concept broadens the framework for addressing agricultural innovation in several ways (Hall *et al.*, 2003; World Bank, 2006). First, and contrary to the linear paradigm, it does not focus narrowly nor primarily on the supply side of innovation (technology pull) or on the demand side (technology push), but rather considers initiatives originating from all actors and stakeholders, be they individuals or institutions, public or private sector, designers or users. The AIS framework also takes into account the knowledge, skills and attitudes of all these actors, be they positive contributions to innovation, or at times resistance against it. It also pays strong attention to the interactions among actors, including brokering, coordination and governance aspects. The market is duly acknowledged as a very potent force for triggering and driving agricultural innovation, as innovative initiatives often emerge from a new market opportunity or signal. Last but not least, the AIS concept highlights the crucial relationships between innovation processes and the socio-economic environment or context in which they are embedded, including the all-important policy aspects.

3.2.3 The agricultural innovation system concept and the World Bank

For its part, the World Bank has gradually embraced the AIS concept. Throughout the 1970s and up to a not-so-distant past, the World Bank invested heavily in building and strengthening National Agricultural Research Systems (Byerlee and Alex, 1994) and strengthening extension through the implementation of the Training and Visit approach to technology transfer (Benor and Baxter, 1984; Hussein *et al.*, 1994). Research and extension were in effect considered at the time the two essential pillars for successful agricultural development, in agreement with the prevailing paradigm of technology development and transfer. The gradual recognition of the existence of a wider knowledge and innovation landscape, reflected in the Agricultural Knowledge and Information Systems (AKIS) approach (Röling, 1990), and eventually the foundation work that led to the 'Enhancing Agricultural innovation' book (World Bank, 2006) contributed to the development and endorsement of the IS concept. The AIS concept has thereafter been consolidated via a major collective undertaking in a recently released 'Agricultural Innovation Systems: An Investment Sourcebook' (World Bank, 2012). The AIS Sourcebook draws on global experience and provides theoretical background combined with evidence-based guidance and examples for AIS investments and initiatives in different agricultural and innovation contexts. The book tackles key AIS elements such as policy coordination and collective action, education and training, extension and advisory services, research systems, engagement of the private sector, enabling environment for innovation, and assessing, monitoring and evaluating AIS investments.

3.3 Operationalizing the innovation system perspective

One reason the AIS perspective is gaining ground rapidly (see EUSCAR, 2012 for a recent review of what has happened throughout Europe) may relate to the fact that it has allowed various lines of works and disciplines to converge into a common framework with the ability to contribute potentially both to diagnosing existing innovation processes and systems and to acting meaningfully upon them. This operationalization, we argue, relies on applying a series of inter-linked approaches and instruments which in their respective ways may contribute to a satisfactory performance of existing and future innovation systems. These approaches and instruments offer concrete ways of achieving such critical objectives as strengthening and facilitating interactions between research and other actors, integrating innovations with markets, developing public-private partnerships, coordinating multiple stakeholders, funding innovation, or transferring technology. They are often used in combination one with another within a given IS initiative. Furthermore, it is illusory to establish a clear hierarchy among them: while approaches sound more generic, in many cases, a so-called 'instrument' (such as a competitive grant scheme) might actually correspond to a higher level goal than an approach (such as the value-chain approach). Several of these approaches and instruments are briefly introduced and illustrated in this section.

3.3.1 Strengthening and facilitating interactions between research and other actors

While in an AIS perspective, research is only one of the multiple actors of a diverse innovation scene, it remains a key actor of the innovation process. Efforts to improve research performance, relevance and contribution to innovation development have started already several decades ago, and are still on-going to this day even though they had to adapt to an environment vastly different from what it used to be.

Developed mostly in the 1980s and 1990s, the various brands of Farming system research (FSR) (Norman and Collinson, 1985; Jouve and Mercoiret, 1987) and Participatory research (PR) (Chambers *et al.*, 1989; Ashby and Sperling, 1995) have helped focus on farmers as primary actors of agricultural innovation. In such approaches, farmers are considered in their diversity in terms of their basic characteristics, needs and also in terms of the environment in which they operate. PR and FSR have made major contributions in popularizing the application of a systemic framework/approach to agriculture at various scales (as reflected in the concepts of cropping, farming and agrarian systems), of which the IS concept is both a direct follow-up and an extension. Recently, these approaches have been broadened in scope and blended with principles of action-research (see below) to form an array of approaches coming under the emerging 'co-design' label (e.g. Béguin, 2003; Triomphe, 2012). Co-design fits particularly well within an AIS framework as it acknowledges and emphasizes both the broad diversity of actors (or stakeholders) with which research collaborates, and the intrinsically iterative, non-linear nature of innovation development.

For their part, development practitioners themselves have proposed similar approaches such as Participatory Technology Development and more recently Participatory Innovation Development (e.g. Hagmman *et al.*, 1996; Veldhuizen *et al.*, 1997; Biggs and Smith, 1998). Close cousins to PR, FSR and co-design, they are much more focused on how to organize and nurture concretely the process of technology/innovation development, giving in doing so primacy not to research but to farmers and other rural stakeholders, as well as to development organizations, including NGOs.

Action-research for its part is a relatively old approach with origins outside of the agricultural field (e.g. Lewin, 1946; Liu, 1997). It has however been recently adapted for use in agriculture (cf. Albaladedjo and Casabianca, 1997; Faure *et al.*, 2010). Action-research provides a valuable framework and generic guidelines for articulating functionally knowledge production (by research but also by users working as lay researchers) and concrete solving of problems faced by, or demands for change made by, users. It also puts forward the need for developing a highly flexible, iterative, reflexive approach to innovation development, which puts a premium on joint planning and learning among the various parties involved. Ensuring effective coordination and governance among diverse stakeholders is also a major issue action-research addresses. It furthermore highlights the role of a shared ethics and values in shaping successful collective action. Last but not least, action-research is very much concerned with the development of new institutions and policies, and not only new technology, as one of the major outputs of the action-research process.

While changing the way research is being conducted is critical for applying an AIS perspective in practice, there are many situations, especially in developing countries, where research and other public or even private actors capable of and willing to invest in innovation development are minimally present at best. Under such conditions, applying an AIS perspective may require that the few existing support institutions (including NGOs) dedicate significant efforts to stimulate local innovation and whenever possible engage in farmer-led joint research. Box 3.1 illustrates the approach to this effect proposed by the PROLINNOVA network across many developing countries.

3.3.2 Improving integration and access to markets

Unsurprisingly perhaps, given that agriculture becomes organically embedded in an increasingly urbanized and globalized society, many recent and on-going innovations in the field of agriculture are being developed with a view to foster, or take advantage of, enhanced market integration, or simply better access to markets. A major objective is to link production more functionally to end-user and consumer demands.

The experience of the Papa Andina programme (Box 3.2), based on the Participatory Market Chain Approach (Devaux *et al.*, 2009), offers a prime example on how market-oriented innovation processes can be facilitated and implemented.

The market being referred to in such approaches can be local, but increasingly, as a result of urbanization and globalization, it is also national (e.g. Papa Andina) and international. Generally speaking, value-chain approaches address the diversity of actors needed to ensure that producers, processors, intermediaries and consumers interact effectively and solve any discrepancies (whether linked to supply or demand) along a value chain, and specifically through the market. Similarly to other approaches involving multiple stakeholders, many value chain approaches also deal with platform development, facilitation and organization of the stakeholders with the help of neutral brokers, and capacity-building with the use of matching grants (see below). The longer-term institutionalization/sustainability of these approaches is, however, often a challenge.

Interestingly, fair access to markets for small holders is rapidly becoming a key concern in many on-going IS initiatives throughout developing countries but also in developed countries. This reflects the importance many countries and actors, public or private, have

Box 3.1. *Stimulating local innovation and farmer-led joint research.*

In an IS perspective, initiatives for innovation may come from any stakeholder. Yet the critical role of farmers as innovators and co-producers of knowledge in their own right is too often overlooked. Prolinnova is an international learning and advocacy network that believes in the change that farmers, both women and men, can make as key decision-makers in sustainable agriculture and natural resource management. Prolinnova strives to develop partnerships and approaches in which the process of innovation development starts with new ideas that farmers have developed or are in the process of developing.

Currently, Prolinnova is made up of networks in 20 countries across Asia, Africa and Latin America. Activities at the country and regional levels are supported by the International Institute of Rural Reconstruction (IIRR), ETC and IED-Afrique. Prolinnova members come from government research and extension agencies, non-governmental organizations (NGOs), educational institutions, community-based/farmer organisations and the private sector.

National network members are encouraged to recognize, document and create awareness about local innovators and local innovations which often go unnoticed and untapped by outsiders. They also develop partnerships with local farmers and their organizations to support on-going processes of local innovation. This includes conducting farmer-led joint experimentation to refine local innovations, exchanging and scaling up of experiences (through innovation fairs, mass media, etc.) and enhancing the capacities of all parties involved in using relevant participatory approaches. One of the critical ways in which farmers can have a say in agricultural research and innovation is to provide them with access to decentralized innovation funds for this purpose. Prolinnova partners are piloting such de-centralised funding mechanisms called Local Innovation Support Funds in eight countries. Initial findings show the usefulness and potential of such schemes for stimulating local innovation and farmer-led joint research.

Network members share lessons periodically at the national and international levels and build on their experiences to promote and support the innovative capacity of small-scale farmers. They also engage actively in policy dialogue at various levels to stimulate and enhance local innovation and create enabling policies for participatory innovation development (Prolinnova website at: www.prolinnova.net; Wongtschowski *et al.*, 2010; World Bank, 2012).

gradually started to attach to small scale family agriculture as a vital contributor to a healthy long-term economic development and to sustainable use of natural resources, of which farmers are both primary users and guardians. This evolution also recognizes that small holders are increasingly producing in response to a market-transmitted consumer demand for production or transformed produce, along with the corresponding norms and quality considerations.

Box 3.2. *Papa Andina and the participatory market chain approach.*

In an IS perspective, market access is often essential for successful innovation. This is reflected in this statement made by Victoriano Meza, a farmer from Pomamanta, a rural community in highland Peru, who says that *'selling my native potatoes to the industry has changed our life.'* He is one of hundreds of farmers benefitting from a new boom in the market for native potatoes. For Mr. Meza, this has meant enough additional income to build a house for his family and equip it with satellite internet *'so that my children can learn quickly and have a better future'*.

This boom stems from the application of the participatory market chain approach (PMCA) approach, within the framework of the Papa Andina programme (a Swiss-funded partnership programme funded in 1998 to promote pro-poor innovation in market chains, with national partners in Bolivia, Ecuador and Peru). One of the aims of Papa Andina was to improve linkages in the potato value chain and particularly to help poor farmers access markets, along with the opportunities markets provide for improving their livelihoods. A first round of PMCA started in 2002 with the objective of developing new potato products. A second round of the PMCA focused on native potatoes (landraces), which are grown mainly by small highland farmers in remote parts of the Andean highlands. In both cases, facilitated multi-stakeholder working groups were formed that analyzed market opportunities and worked to develop new products.

Key outcomes of the two rounds of PMCA include the creation of a new brand of high-quality fresh potatoes for the wholesale market, a new native potato chip product and brand, and the first brand of high-quality native potatoes to be marketed in Peruvian supermarkets. There were also technological innovations in the form of improved pest and disease management or selection of harvested produce. Also, a national-level platform, CAPAC-Peru, was established to promote marketing of quality potato products and promote innovation, in which local actors are gradually taking increasing responsibility as their capacity and trust builds. CAPAC helped organize small farmers supply potatoes that meet the more demanding market requirements. Soon thereafter, other actors started to realize the untapped market potential of native potatoes in urban markets and created new products based on them. Today, high-quality native potatoes and potato chips are sold under various brand names in most of Peru's leading supermarkets. With the entry of a multi-national company into the market, Papa Andina began to work on corporate social responsibility, which has helped balance corporate interests with those of community suppliers and the environment. Last but not least, an annual National Potato Day has been established in Peru.

These many innovations and developments have contributed to increased farmer income and self-esteem; more stable markets; better organized farmers; better coordination among stakeholders and the popularization of native potatoes in Peru's urban cuisine.

Some challenges remain and are currently being addressed: for example, coordination and facilitation, which requires full-time facilitators/brokers, requires substantial investment in capacity-building; financial sustainability beyond donor funding. Also, traditional evaluation approaches do not fit well with learning-based innovation processes (Adapted from Devaux *et al.*, 2009; World Bank, 2012).

3.3.3 Developing public-private partnerships

Since the nineties, emphasis has been given by many international and national donors, including the World Bank and FAO, to the establishment of more systematic public-private partnerships (PPPs). In a PPP, at least one public and one private organization share resources, knowledge, and risks to achieve a match of interests and jointly deliver novel products and services. The objective of PPPs is to enable more effective relationships between, on the one hand, public actors of the ARD system, and, on the other, private sector actors, which can complement the public investments and bring to innovation the much needed entrepreneurial skills and link to markets. While PPPs have often been used until now to deliver social and environmental services (such as providing and treating drinking water), they are increasingly applied to deliver innovation and technology adoption (such as the production by Kenyan small holders of sorghum for a private brewery, or producing pig meat from an endangered species for a Chinese pig-breeding and processing company). PPPs may help overcome problems of underinvestment in agricultural innovation and accelerate technological progress by lowering the risk for individual private partners while at the same time fostering the adaptation and dissemination of new research results or existing knowledge and technologies. Quite a number of agricultural sector PPPs have been set up over the last ten or more years in developing countries (Hartwich and Tola, 2007; Spielman *et al.*, 2009). Successful examples of PPPs include cases where research has helped to reduce the costs of processing primary products, cases in which improving product quality helped to access higher-value markets, or cases of exchanges of planting material and outsourcing seed multiplication to the private sector (World Bank, 2012). However, a healthy balance between private commercial interests and the interests of society requires paying attention to having clear criteria in place to determine when public intervention is justified and at what level, as well as guidelines, and a clear time frame and exit strategy (World Bank, 2012).

As other approaches, PPPs do not operate in a vacuum: they are often associated with diverse instruments that provide incentives for partnering (grants, incubators, consortiums, PMCA) and approaches that address transaction costs (brokering arrangements, platforms) and intellectual property rights (licensing agreements, technology transfer offices).

3.3.4 Creating and strengthening multi-stakeholder innovation platforms, networks and consortiums

In most approaches to innovation involving diverse stakeholders, significant efforts are geared at creating or strengthening functional multi-stakeholder innovation platforms, networks or consortiums. Platforms and networks are seen as a way of facilitating the needed interactions among stakeholders and of ensuring their effective coordination by decreasing otherwise inherently high transaction costs (Nederlof *et al.*, 2011; Hounkonnou *et al.*, 2012).

Consortiums for their part can be seen as a particular case of formalized multi-stakeholder platforms that bring together diverse public and private partners and users around specific and common problems requiring R&D investment. Consortium members jointly define the corresponding R&D strategies, and finance and implement the subsequent activities. Most consortiums have a lead organization, and each partner has a specific role and commits resources. Contributions from a range of actors, including private enterprises, cover various aspects of R&D (demand identification, R&D investment, technology transfer and adoption). Consortiums are often funded through competitive grants for a limited period (World Bank, 2012). Consortium approaches have been applied, e.g. in Australia, Chile and India, where they have been able to bring in diverse actors, including private sector, needed for innovation to take place – covering the entire R&D continuum (See Box 3.3: NAIP case).

Whether in the case of multi-stakeholder platforms, networks or consortiums, a recurring issue is to go beyond self-organization of such platforms or networks to ensure smooth, continuous interaction and collaboration among stakeholders by developing or fulfilling brokering functions (Perez Perdomo *et al.*, 2010), which involve leading, coaxing, catalyzing, and championing the set of stakeholders in the face of obstacles or conflict (see also Box 3.3 for the role played by the Help Desk in the NAIP experience) and providing support for addressing benefit sharing issues (e.g. IPR). In response to what is increasingly recognized as a critical gap in the typical structure and functioning of innovation systems, new brokering institutions may actually emerge to fill it as their core mandate, as is the case in the Netherlands (e.g. Klerkx and Leeuwis, 2008).

3.3.5 Funding innovation through grants and other novel mechanisms

The initiatives and approaches described above often require access to funding mechanisms that can help the involved stakeholders support the various costs and expenses involved.

Among the diverse funding mechanisms (e.g. grants, guarantees, risk capital, tax incentives), grants, and, in particular, competitive research grants, have been a commonly applied instrument to promote diverse innovation activities, such as demand-driven research, adaptive research, research-extension-farmer linkages that improve the relevance and dissemination of new technologies, demand-driven services, productive partnerships, and links to markets (World Bank, 2010).

While much attention has been given until now to providing funding for research institutions and to a lesser extent to extension and farmers, less attention has been paid to engaging other actors and farmers in innovation approaches (World Bank, 2010). Increasingly however, IS initiatives such as NAIP or Papa Andina have benefitted from the emergence of new funding mechanisms featuring improved access to innovation resources for a wider set of actors (to the point that some grant schemes require diverse stakeholders to submit a proposal together and propose to share the grant 'fairly' among them). Matching grants (in which grant funding

Box 3.3. Indian national agricultural innovation project.

The National Agricultural Innovation Project (NAIP), launched in 2006, addresses R&D and innovation challenges by changing the way in which scientists, farmers and agricultural entrepreneurs interact in the national AIS. This project funded by the World Bank sought to strengthen the role of the Indian Council for Agricultural Research (ICAR) in catalyzing and managing change in the NARS and to promote the development of three kinds of multi-stakeholder, multidisciplinary consortiums involving public and private organizations, universities, NGOs, and others: (1) market-oriented, collaborative research consortiums with a focus on selected agricultural value chains; (2) livelihood research consortiums with a focus on strategies to sustain secure rural livelihoods in about 110 disadvantaged districts; and (3) basic and strategic research consortiums with a focus on well-defined areas of frontier science with potential applications for problems in Indian agriculture.

Promising consortiums and research alliances were funded by NAIP through a competitive process. In each case, consortium members were jointly responsible for the governance, design, and implementation of their research programmes; maintaining satisfactory fiduciary and safeguard arrangements; applying the resulting innovations; and disseminating new knowledge through conferences, innovation marketplaces, networks, and communications strategies.

A Helpdesk was established to support the new and more challenging partnerships that the consortiums represented. It provided guidance for preparing concept notes and full research proposals, assisted in matching consortium partners and helped to overcome initial problems in managing the consortiums. In doing so, it made use of a number of tools: the Helpdesk portal, e-learning and multimedia modules, databases of potential partner institutions and organizations, case studies of agricultural projects using a consortium approach, frequently asked questions (FAQs), meetings, workshops, reviews and direct email responses to potential consortium members.

The outcomes of NAIP are numerous and diverse. The project/approach received an overwhelming amount of interest and resulted in funding of 188 consortiums. NAIP was able to introduce greater pluralism into agricultural research, with almost 40% of consortium institutes coming from outside the ICAR-state agricultural university system. PPPs were promoted on a large scale for the first time. The consortium approach has promoted synergy, teamwork, partnership, value addition, learning, enhanced focus on high priority topics and better, more relevant research. A greater impact is anticipated thanks to new technologies and products developed with the context of the NAIP-induced partnerships. Lastly, the institutions taking part in the consortiums have been strengthened as a result of formal training and through development of new multi-stakeholder partnerships. This is illustrated by the continuous interaction between public, private, and NGO sectors and the willingness of ICAR institutes to work outside their usual sphere. ICAR has started to mainstream the consortium approach and competitive selection process throughout its institutes.

Challenges include effective partnership coordination, building staff commitment and partnering skills; learning costs for the new actors/partners coming on-board; and effective monitoring and evaluation procedures (World Bank, 2006b, 2012).

matches contributions made by a set of stakeholders with their own resources), often in the form of national innovation funds, have gained ground as they are particularly well-suited for promoting near-market technology generation, technology transfer and adoption, private economic activity, and overall innovation. By bringing further attention to demand and use from the very beginning, basically by attracting users of technologies and knowledge in partnerships, matching grants may be more effective than competitive research grants at enhancing the use of technology and knowledge by farmers and other entrepreneurs (World Bank, 2010).

The question of matching grants may however be problematic in situations where many stakeholders are resource-poor. Recently, funding mechanisms targeting farmers and aiming at supporting them for innovation have been developed, cf. experience of the Prolinnova network with the Farmer Access to Innovation Resources programme (Wongtschowski *et al,* 2010, and also Box 3.1). While this is not automatically the case, grants to farmers may include provisions for ensuring other stakeholders (researchers, extension, etc.) are called in as service providers (Word Bank, 2010; Triomphe et al., 2012).

3.3.6 Promoting technology commercialization through technology transfer offices, incubators and science parks

While the AIS perspective does not consider technology transfer as such as sufficient for spurring innovation, the capacity to manage formal technology transfer mechanisms remains critical in many situations, such as to engage effectively in PPPs and, increasingly, to transfer technologies that can be disseminated through market channels.

Technology transfer offices (TTOs) are special units affiliated with a research organization or university with a mandate to identify and protect as well as facilitate the use and commercialization of research results. Very common in developed countries, such units are increasingly created in developing countries as a way of diversifying funding sources, engaging public sector staff in commercial activities and commercializing technology. TTOs can expand the recognition of the research organization's work (thereby strengthening public perceptions of its value), move technologies to end-users (e.g. seed companies, farmers), and generate revenues to fund continuous research. TTOs can provide special expertise on Intellectual Property protection and/or legal agreements and contribute to formal transfers of technology from public organizations or universities or from the private sector to commercial or international partners.

Some technology transfer offices also host *incubators* to help technology-oriented firms (often established by researchers) commercialize new technology. Creating strong, organic linkages between research and enterprises, and providing them with an enabling environment capable of fostering and nurturing their emergence is at the heart of the incubator approach. Incubators provide hands-on management assistance, access to financing, business and technical support services, shared office space, and access to equipment (World Bank, 2012). For example, ICRISAT has established an incubator that commercializes small-holder focused technologies, developed by Indian researchers. Most incubators however do have a wider mandate to support innovation by small and medium enterprises and to promote overall agribusiness development.

Science parks (also called technology or research parks) are, for their part, organizations whose main aim is to increase local wealth by promoting a culture of innovation and improving the competitiveness of local businesses and knowledge-based institutions. They do so through stimulating and managing the flow of knowledge and technology among involved stakeholders, through incubation and spin-off processes and by providing other value-added services together with high-quality space and facilities. They are useful for fostering PPPs in more mature innovation contexts. They function best where there is investment capital from the private sector, industrial engineering expertise, and a sufficient knowledge and technology base. They are a useful nexus between the private sector and research institutes (particularly universities), taking promising research products to market and providing backstopping for product modification. Their diverse services include facilitating the creation of public-private partnerships for research, providing infrastructure, and providing other services, including business development (World Bank, 2012).

3.4 Some key challenges in applying an AIS perspective in practice

While the implementation of the approaches and instruments described above has already yielded very valuable contributions to innovation development, they also meet significant challenges.

The sheer complexity in understanding, assessing and, perhaps more importantly, designing and steering IS setups is one key challenge. As mentioned by CTA at ISDA, '*It is difficult for most scientists to grasp the innovation system concept which is conceptually diffuse, rooted in economic theory and far removed from their specific disciplinary training*.' Even the so-called 'innovation brokers' or 'innovation coaches' with years of experience may still suffer from a lack of clear basic understanding of Innovation Systems thinking (Pyburn and Woodhill, 2011). This is because the IS field is largely couched in rather abstract, often vague social science language and notions (networks, platforms, institutional innovation, to name but a few), which are rather difficult to communicate and require a significant theoretical background as an entry point. Even if one understands the concepts, one is then faced with the actual operationalization of innovation systems thinking and perspective, which to this day remains limited.

Another challenge is the unclear, hard-to-prove causality between, on the one hand, the use of a given set of potentially useful IS approaches or instruments (such as the ones described in the section above) and, on the other, achieving a particular positive innovation outcome.

In other words, the success and direction of an innovation process does not only depend on the approaches used. A lot seems to depend on factors and conditions pertaining to the rather fuzzy 'enabling environment', which varies tremendously, e.g. between more agrarian vs. more market-oriented contexts, and also within a given country (World Bank, 2012). In particular, a lot depends on the supportive and usually complex policy and fiscal framework for science, technology, legal, advisory, and trade issues, which shape innovation in complex direct and indirect ways.

AIS approaches take a strong view about changing the way innovation should be tackled. However, scaling up and institutionalization not only of results (the famous yet elusive 'transfer of technology') but of novel approaches to innovation is particularly challenging, as it may require changing the way entire institutions and bureaucracies usually work, which takes time. To do so, it is necessary not only to invest in learning and capacity building (see below), but to provide incentives that allow actors to put new skills into use, to change unsuitable regulatory or policy frameworks. There is also a need to nurture new attitudes and practices, such as openness, flexibility and capacity of adaptation. In short, scaling up and institutionalization require changing part of the enabling environment mentioned above.

Such far-reaching changes in turn require the creation and strengthening of enough human capital to meet the needs of the various stakeholders. Some key principles have been identified for doing so: (1) develop new educational programmes that are more strategically attuned to the needs of social and productive actors; (2) develop new curricula that instil the capacity to deal with complexity, change, and multi-actor processes in rural innovation, in addition to allowing greater specialization; and (3) strengthen the innovative capabilities of organizations and professionals involved in agricultural education and training. However, the corresponding investments in capacity-building are costly and cannot take place overnight. Also, some stakeholders, especially the poorest ones, might not have easy access to capacity-building services.

While few authors address this dimension explicitly, applying an AIS perspective brings to the fore the challenges of dealing concretely with marked power, resource and capacity asymmetries among stakeholders, which often have underlying political dimensions as well. Such asymmetries are a staple of innovation setups in most developing countries (Hocdé *et al.*, 2009; Faure *et al.*, 2010). How much IS initiatives may concretely contribute to reduce asymmetries by, for example, strengthening the voices, capacities and autonomy of smallholders and other weaker rural actors is a matter of much debate. The Prolinnova (see Box 3.1) and Papa Andina (see Box 3.2) initiatives clearly have a pro-poor agenda in fostering innovation. Despite such encouraging examples, there is however nothing automatic nor easy about ensuring that such weak stakeholders are able to interact meaningfully and fairly with more powerful and better-organized ones, and in doing so, may actually benefit from innovation dynamics. The 'establishment' in institutions of agricultural research, extension and education – often the middle-level staff in government organizations – tends to resent and resist sharing their power and resources, for fear of losing control and because they are uncertain about their new roles. But farmers also find it difficult to learn new roles. Effective brokering arrangements and capacity-building of the weaker stakeholders may play a positive role in mitigating asymmetries. However, more often than not, the resolution of such problems, if and when feasible, depends on long-term evolutions and interventions usually way beyond the scope of IS initiatives. This in turn questions the time horizon of many innovation interventions, especially those taking place within the framework of externally-funded projects over short periods of time, whose transitory positive effects obtained under artificial conditions of resources and rules of the game tend to falter as soon as the projects end.

Last but not least, one must mention the challenges related to a proper monitoring and evaluation of the impact of AIS approaches, instruments and investments (EU SCAR, 2012). Given the novelty of the AIS approaches, and the expectations about their ability to deliver better and more relevant innovation, being able to assess their actual impact and to compare IS initiatives one with another would appear essential. Yet many on-going IS experiences highlight the difficulty of the task at hand, as dimensions to monitor are many, and indicators needed to characterize them properly over time and space and over the diversity of stakeholders and activities involved would require the estimation of a host of quantitative and qualitative variables whose meaning and interpretation might not be straightforward. Cost of such monitoring is also an issue, as few stakeholders or funding agencies are willing or simply able to fund a thorough and consequently rather costly monitoring and evaluation system.

3.5 Conclusions and perspectives

In this brief overview, we have discussed the meaning and value of the emerging Innovation Systems perspective, along with some of the diverse approaches and instruments with which it can be implemented in practice in the agricultural sector, and some of the many challenges it faces. Current AIS approaches build clearly upon the valuable lessons learnt with previous approaches and paradigms, such as the strengthening of NARS and AKIS. Basic investments in R&D infrastructure and equipment, in human resources, and in system performance and accountability remain critical elementary building blocks on which an AIS perspective brings complementary facets, such as an emphasis on diverse stakeholders and their interactions, on skills and linkages with markets, while also paying close attention to the wider policy and enabling environment.

It is still early to assess the full impact of the AIS approach in nurturing innovation and contributing to large scale impact, especially if one wishes to compare the value of the AIS perspective to that of other historical and still largely prevalent paradigms such as the linear model of technology development and transfer. However, the intrinsic attractiveness of AIS is strong, and increasingly, actors and organizations, institutions and policy-makers in developed and developing countries alike are investing in projects and initiatives at a larger scale in an attempt to bolster innovation.

Clearly, there will be challenges along the way, and the AIS perspective, whatever its intrinsic potential, is no silver bullet or panacea to ensuring that the ever changing and evasive need for innovation in agriculture and rural development will be answered in a satisfactory manner in all contexts.

At this stage, efforts are needed in several directions to make sure the AIS approach and perspective reaches its full potential and brings in the expected benefits:

- Share further the potential of the AIS approach with decision-makers and investors. While the AIS Sourcebook (World Bank, 2012) represents one such endeavour for the World Bank and its partners, similar efforts based on more diverse and targeted evidence might be needed for communicating the rationale and presenting convincing evidence from well-chosen pilot experiences on the impact of an AIS approach to other decision-makers and investors worldwide. One of the aims will be to stimulate political will and to provide relevant (economic) arguments to ministries, public institutions and the private sector alike to endorse and adopt an effective IS approach.
- Develop appropriate and well-resourced innovation financing mechanisms. Learning from past experiences, such mechanisms need to take into account the multiple dimensions of AIS approaches and the underlying dynamics of real-like innovation processes. They must allow diverse actors to get flexible access to the resources they need to be part and parcel of vibrant innovations scenes at scales relevant to them, in tune with the evolving needs of the users and responsive to the incessant changes affecting the overall or local environment.
- Establish a community of practice on AIS. AIS efforts are still few and fragmented, and hence the pace of progress in understanding AIS, and learning from past or on-going AIS initiatives is still relatively slow. With leading agricultural universities eventually catching up with teaching IS concepts to new generations of agricultural professionals, establishing a vibrant community of practice around AIS could be a worthwhile way of accelerating collective learning and expertise about AIS experiences, and of further developing the AIS approaches and instruments needed for its successful implementation in diverse contexts;
- Further improve the understanding of the AIS perspective, its implications and impact. Although the AIS perspective is gradually moving toward an operational framework, it remains an evolving concept and practice that requires better understanding and analysis. More attention must also be paid to a proper assessment of IS initiatives and their impact, which in turn requires the development of specific and suitable capacities, tools and methods, as well as indicators which could allow meaningful and lessons-rich comparisons between AIS experiences at the national or international levels.

To a greater or lesser extent, many of the efforts mentioned above are under way. They are gradually providing the necessary inputs for a rigorous, reflective, flexible and context-conscious application of the AIS approach. In doing so, constant efforts must be made to keep the AIS perspective and its suite of approaches and instruments flexible, diverse and adaptable enough to the specific environments in which one wishes to apply it. Failing to do and turning the AIS perspective into a rigid, mechanical, 'one size fits all' blueprint for making innovation happen would be a sure recipe for future disillusionments.

References

Akrich, M., Callon, M. and Latour, B., 1988a. A quoi tient le succès des innovations. Premier épisode: l'art de l'intéressement. Gérer et comprendre, Annales des Mines, 1988, 11, 4-17.

Akrich, M., Callon, M. and Latour, B., 1988b. A quoi tient le succès des innovations. Deuxième épisode: l'art de choisir les bons porte-parole. Gérer et comprendre, Annales des Mines, 1988, 12, 14-29.

Albaladejo, C. and Casabianca, F. (eds.), 1997. La recherche-action. Ambitions, pratiques, débats. INRA – SAD, Paris. Etudes et Recherches sur les systèmes Agraires et le Développement no.30.

Alter, N., 2000. La trajectoire des innovations. In: Alter, N. (ed.) L'innovation ordinaire. PUF, Paris, France, pp. 7-39.

Ashby, J. and Sperling, L., 1995. Institutionalizing participatory, client-driven research and technology development in agriculture. Development and Change, 26, 753-770.

Benor, D. and Baxter, M., 1984. Training and visit extension. World Bank and International Bank for Reconstruction and Development, Washington, DC, USA.

Béguin, P., 2003. Design as a mutual learning process between users and designers. Interacting with Computers, 15 (5), 709-730.

Biggs, S. and Smith, G., 1998. Beyond methodologies: coalition-building for participatory technology development. World Development, 26(2), 239-248.

Byerlee, D. and Alex, G.E., 1998. Strengthening national agricultural research systems. policy issues and good practice. World Bank, Washington, DC, USA.

Callon, M., Lascoumes, P. and Barthe, Y., 2001. Acting in an uncertain world. An essay on technical democracy. The Mit Press, Cambridge, MA, USA.

Chambers, R., Pacey, A. and Thrupp, L.A., 1989. Farmer first. Farmer innovation and agricultural research. Intermediate Technology Publication, London, UK.

Devaux, A., Horton, D., Velasco, C., Thiele, G., Lopez, G., Bernet, T., Reinoso, I. and Ordinola M., 2009. Collective action for market chain innovation in the Andes. Food Policy, 34(1), 31-38.

EU SCAR, 2012. Agricultural knowledge and innovations systems in transition. A reflection paper. Standing Committee on Agricultural Research, Brussels, Belgium.

Evenson, R.E. and Gollin, D., 2003. Assessing the impact of the green revolution, 1960 to 2000. Science, 300, 758-762.

Faure, G., Gasselin, P., Triomphe, B., Temple, L. and Hocdé, H., 2010. Innover avec les acteurs du monde rural. La recherche-action en partenariat. Collection 'Agricultures tropicales en poche'. QUAE, CTA, Gembloux, France.

Freeman, C., 1988. Japan: a new national system of innovation? Technical change and economic theory. Pinter, London, UK.

Fukuda, C. andWatanabe, C., 2008. Japanese and US perspectives on the national innovation ecosystem. Technology in Society, 30(1), 49-63.

Hocdé, H., Triomphe, B., Faure, G. and Dulcire, M., 2009. From participation to partnership, a different way for researchers to accompany innovations processes: challenges and difficulties. In: Sanginga, P., Waters-Bayer, A., Kaaria, S., Njuki, J. and Wettasinha C (eds.). Innovation Africa: enriching farmers' livelihoods. Earthscan, London, UK, pp. 135-150.

Hagmman, J., Chuma, E. and Murwira, K., 1996. Improving the output of agricultural extension and research through participatory innovation development and extension; experiences from Zimbabwe. European Journal of Agricultural Education and Extension, 2 (4), 15-23.

Hall, A., Sulaiman, V.R., Clark, N. and Yogoband, B., 2003. From measuring impact to learning institutional lessons: an innovation systems perspective on improving the management of international agricultural research. Agricultural Systems, 78, 213-241.

Hartwich, F. and Tola, J., 2007. Public-private partnerships for agricultural innovation: concepts and experiences from 124 cases in Latin America. International Journal of Agricultural Resources, Governance and Ecology, 6(2), 240-255.

Hounkonnou, D., Kossou, D., Kuyper, T.W., Leeuwis, C., Nederlof, S., Röling, N., Sakyi-Dawson, O., Traoré, M. and Van Huis, A., 2012. An innovation systems approach to institutional change: Smallholder development in West Africa. Agricultural Systems, 108, 74-83.

Hussain, S. S., Byerlee, D. and Heisey, P.W., 1994. Impacts of the training and visit extension system on farmers' knowledge and adoption of technology: evidence from Pakistan. Agricultural Economics, 10(1), 39-47.

Jouve, P. and Mercoiret, M.-R., 1987. La recherche développement: une démarche pour mettre les recherches sur les systèmes de production au service du développement rural in Les Cahiers de la Recherche-Développement, (16), 8-13.

Klerkx, L. and Leeuwis, C., 2008. Matching demand and supply in the agricultural knowledge infrastructure: experiences with innovation intermediaries. Food Policy, 33, 260-276.

Knickel, K., Tisenkopfs, T. and Peter, S. (eds.), 2009. Innovation processes in agriculture and rural development. Results of a cross-national analysis of the situation in seven countries, research gaps and recommendations. In-Sight project report.

Latour, B. 1987. Science in action: how to follow scientists and engineers through society. Open University Press, Milton Keynes, UK.

Lewin, K., 1946. Action research and minority problems. Journal of Social Issues, 2, 34-46.

Liu, M., 1997. Fondements et pratiques de la recherche-action. L'Harmattan, Paris, France.

Lundvall, B.A., 1992. National systems of innovation. Pinter, London, UK.

Metcalfe, J.S., 1995. The economic foundations of technology policy. In: Stoneman, P. (ed.): Handbook of the economics of innovation and technological change. Oxford University Press, Oxford, UK, pp. 409-512.

Nederlof, S., Wongtschowski, M. and Van der Lee, F. (eds.), 2011. Putting heads together. Agricultural innovation platforms in practice. Bulletin 396, KIT Publishers, Amsterdam, the Netherlands.

Nelson, R., 1993. National innovation systems: a comparative analysis. University Press, Oxford, UK.

Norman, D. and Collinson, M., 1985. Farming systems research in theory and practice. In: Agricultural Systems research for developing countries, Richmond, Australia, 12-15 May 1985, ACIAR, no. 11, pp. 16-30.

OECD, 1997. National innovation systems. Organization of Economic Cooperation and Development, Paris, France.

Perez Perdomo, S., Klerkx, L. and Leeuwis, C., 2010. Innovation brokers and their roles in value chain-network innovation: preliminary findings and a research agenda. In: Innovation and Sustainable Development in Agriculture, Symposium Proceedings, June 28-30, 2010, Montpellier, France.

Pyburn, R. and Woodhill, J. (eds.), 2011. Dynamics of rural innovation: a primer for emerging professionals. KIT Publishers, Amsterdam, the netherlands.

Rajalahti, R., Janssen, W. and Pehu, E., 2008. Agricultural Innovation systems: from diagnostics toward operational practices. Discussion Paper 38. World Bank, Washington, DC, USA.

Rajalahti, R., Woelcke, J. and Pehu, E., 2005. Development of research systems to support the changing agricultural sector: proceedings. Agriculture and Rural Development Discussion Paper 14. World Bank, Washington, DC, USA.

Rogers, E.M., 1983. Diffusion of innovations, 3rd ed. Free Press, New York, NY, USA.

Röling, N., 1990. The agricultural research-technology transfer interface: a knowledge systems perspective. In: Kaimowitz, D. (ed.), Making the Link; Agricultural Research and Technology Transfer in Developing Countries. Westview Press, Boulder, Co, USA, pp. 1-4 and 11-23.

Schumpeter, J., 1942. Capitalism, socialism, and democracy. Harper and Row, New York, NY, USA.

Spielman, D.J., Hartwich, F. and Grebmer, K., 2009. Public-private partnerships and developing-country agriculture: Evidence from the international agricultural research system. Public Administration and Development, 30 (4), 261-276.

Triomphe, B., 2012. Codesigning innovations: how can research engage with multiple stakeholders? In: World Bank (ed.) Agricultural Innovation Systems. An Investment Sourcebook. World Bank, Washington, DC, USA, pp. 308-315.

Triomphe, B., Wongtschowski, M., Krone, A., Waters-Bayer, A., Lugg, D. and Van Veldhuizan, L., 2012. Providing farmers with direct access to innovation funds. In: World Bank (ed.) Agricultural innovation systems. An investment sourcebook, World Bank, Washington, DC, USA, pp. 435-441.

Veldhuizen, L., Waters-Bayer, A. and De Zeeuw, H., 1997. Developing technology with farmers. ETC Netherlands and Zed Books, London, UK.

Wongtschowski M., Triomphe B., Krone A., Waters-Bayer A. and Van Veldhuizen L., 2010. Towards a farmer-governed approach to agricultural research for development: lessons from international experiences with local innovation support funds. In: Innovation and Sustainable Development in Agriculture and Food – ISDA 2010, Montpellier, France. Available at: http://hal.archives-ouvertes.fr/hal-00510417/fr/.

World Bank, 2006a. Enhancing agricultural innovation: how to go beyond the strengthening of research systems. World Bank, Washington, DC, USA.

World Bank, 2006b. India: national agricultural innovation project, project appraisal document. Report No. 34908-IN, Agriculture and Rural Development Sector Unit, South Asia Region, Washington, DC, USA.

World Bank, 2010. Designing and implementing agricultural innovation funds: lessons from competitive and matching grant projects. Washington, DC, USA.

World Bank, 2012. Agricultural innovation systems. an investment sourcebook. World Bank, Washington, DC, USA.

Chapter 4. Innovation systems of the future: what sort of entrepreneurs do we need?

Andy Hall and Kumuda Dorai

4.1 Introduction

Agricultural innovation invariably involves a whole range of partnerships, alliances and network-like arrangements that connect together knowledge users, knowledge producers and others involved in enabling innovation in the market, policy and civil society arenas. There is now a very large conceptual and empirical literature that reveals agricultural innovation not as process of invention driven by research, but as a process of making novel use of ideas (old and new) with the specific intention of adding social, economic and/or environmental value (Juma, 2010).

It is only in the last 20 years or so that policy frameworks have gradually come to acknowledge what has been written about in the literature for long: agricultural innovation emerges out of this interaction among a range of actors. This is reflected in the widely accepted convention of conceiving agricultural research not as a stand-alone intervention but as part of an agricultural knowledge system or, more recently, as part of an agricultural innovation system (World Bank, 2006).

What is increasingly becoming significant is that the range of actors involved in such partnerships now goes beyond the traditional domains of public-private sector actors, NGOs and research, and has come to include a whole host of new players who often challenge conventional thinking and blur the lines or straddle multiple roles in pursuing their mandates of social and environmental good while still keeping note of profitability margins.

This new class of entrepreneurs, often operating below the market and policy radar[1], is pioneering new, disruptive modes of innovation that address the social and environmental concerns that public policy is currently struggling to deal with. The emphasis on the private sector in innovation systems thinking – and an assumption that this means companies – has obscured the importance of other forms of entrepreneurship. Many of these have a long history in development practice. Perhaps it is time to look below-the-radar and support the entrepreneurship we find there.

[1] We refer to these enterprises as 'below-the-radar' in terms of being ignored or overlooked, not just by policy-makers but also the mainstream private sector actors, as part of the innovation landscape and having something important to contribute to its further development.

These are troubling times for those who like to classify organizations into comfortably familiar, watertight, categories. Take for instance the Bangladesh Rural Advancement Committee, better known as BRAC. This is a non-governmental organization with a mandate of 'empowering people and communities in situations of poverty, illiteracy, disease and social injustice'. But this is also an NGO that is a major value chain player involved in both inputs supply and agro-processing. It poses serious competition to private agro-industries, yet it services the needs of its poor clientele. It flags social entrepreneurship as one of its key strategies. Or take Amul – the famous dairy cooperative in India, jointly owned by 3.03 million Indian milk producers which collects milk from producers, processes it into various products and markets the milk and milk products in markets in India and abroad. All the while Amul also provides necessary input services to its producers and conducts training to develop skills.

BRAC and Amul are just a couple of many such actors emerging as key players in the development arena that policy largely tends to overlook and which the agricultural development literature is just gradually beginning to acknowledge (see Box 4.1).

Do these sorts of organizations offer new avenues for innovation capacity building at a time when the idea of constructing pro-poor innovation systems is starting to run into the sand? To answer this, it is first useful to go back and look at what the innovation systems idea has had to contribute and where it has got stuck.

4.2 Innovation systems: beyond concepts?

A key feature of the debate about the nature of agricultural innovation in developing countries has been the understanding that it is a process embedded in a much wider set of relationships than those implied by research-extension-farmer linkages. Biggs and Clay (1981) and Biggs (1990) talked of different and multiple sources of innovation. Röling (1992) introduced the idea of agricultural knowledge systems and the Wageningen School of Innovation Studies used such ideas to explore multi-actor rural innovation landscapes (see Engel, 1995; Leeuwis and Pyburn, 2002). At the same time two interlinked debates were also underway. First, starting in the 1980s, was the global trend to revisit the role of the state in national economies. This rolled out from Europe and North America to emerging economies of South Asia and Africa as an accompaniment to development assistance and development bank investment conditionality – so-called structural adjustment. The second was the reappraisal of the role of private sector activity in innovation in the most successful economies and the realisation that innovation was not solely associated with public R&D (or even R&D, more generally) but was an activity distributed through the whole of the economy. The most successful countries were found to be those where dense networks of interaction underpinned a national system of innovation (Freeman, 1995).

Box 4.1. Examples of entrepreneurship.

Real IPM (Kenya). Real IPM is a private company in Kenya that makes its money producing and selling bio-pesticides to the commercial horticultural sector in East Africa. Not only is its philosophy deeply green, but it has also recently moved into a new and large market opportunity – poor farmers. In collaboration with the UK's Department for International Development's (DFID) Research Into Use (RIU) programme, it developed a network of village-based advisers to help promote and sell a bio-pesticide to combat *striga*, a parasitic weed of the food staple crop maize, to small-scale farmers. The company encountered numerous regulatory hurdles in registering its products for commercial application – legislation made provision for licensing chemical pesticides but not bio-control agents. Real IPM was forced to engage in policy brokering activities that included engaging with government officials and scientists, presenting scientific evidence from field trials and building relationships with a range of interested stakeholders in the regulatory process. This brought about a review of import regulations and the consequent drafting of Kenya's regulations for biological inputs, although the bio-control agent is yet to be licensed for sale.

FIPS (Kenya). Farm Input Promotion Services (FIPS) is a company promoting small seed and fertiliser packs among smallholder farmers in Kenya. Adoption of fertiliser and improved seed in parts of rural Kenya is low due to the high cost of large packs of inputs. With limited demand many rural areas are poorly served by private input suppliers. FIPS has tried to address this by kickstarting demand by negotiating with input companies to provide small low-cost seed and fertiliser packs. FIPS then provides technical advice to farmers to encourage them to use these inputs. To achieve this, FIPS brokers networks between farmers and local input stockists, public research institutes and input manufacturers. Providing farmers access to technology has been a starting point, but FIPS is now playing a role in brokering linkages with local supply systems and research.

TerraCycle Inc. (USA). TerraCycle is a private company in the USA that specializes in making consumer products from waste materials that are hard to recycle. Starting out with producing fertiliser from worm waste products (feeding organic waste to worms), the company has diversified its range of products to include bags, park benches and flower pots made from the packaging waste of other companies' products.

Gram Mooligai Company Ltd. (India). Gram Mooligai is a public limited company that specializes in collecting, cultivating, processing and marketing medicinal plants and honey abroad. It is owned through equitable profit sharing by several women's self-help groups, involving over 1000 households, in South India. The company, and the foundation that set it up through donor funds, also spend their resources in training farmers in organic farming practices.

This idea of a national system of innovation was later adapted to explore the innovation process in agricultural development (Hall, 2005; Hall *et al.*, 1998, 2002). This built on the work of Biggs, Röling and others, but was much more explicit about the importance of private sector actors in the agricultural innovation process and flagged the fact that the macro enabling environment for innovation was as important as the micro-innovation activities of farmers in the rural space (World Bank, 2006). Early work rather prematurely predicted that such private sector-dependent models of innovation had potential for development goals such as poverty reduction (Hall *et al.*, 2002). More recent writing on agricultural innovation systems has embraced a greater diversity of innovation arrangements – some more participatory, some more research-driven (Hall, 2009). This line of thinking nevertheless flags the importance of the private sector for different types of innovation activity and at different points in the innovation trajectory (Hall, 2006, 2009).

The greatest contribution of the innovation systems idea to agricultural and rural development is conceptual. It has helped planners reconsider the place of agricultural research within dynamic processes of innovation and it has revealed the importance of linkages needed to connect research to others involved in this process. However, the innovation system idea is not a new blueprint for how to organize innovation, but a metaphor to explain the vast diversity of ways of organizing innovation for different purposes and in different contexts. But how can its central message of distributed (rather than centralized) creativity be taken forward to address social and environmental sustainability aspects of the international development agenda?

A number of operational strategies have been suggested to help strengthen linkages and coherence in different areas of action and policy. Many of these hinge upon the idea of public-private sector partnerships, and, more generally, the role of the private sector as a driver of the innovation process. In practice, partnership building has proven difficult. Even in cases where new alliances have been developed, the real sticking point is the governance of these to direct innovation towards a social and sustainable development agenda. This seems to be the area where innovation systems ideas are reaching their limits as a guide to practice.

4.3 Alternative sources of disruption: entrepreneurship and development

The glaring paradox is that while we have been struggling to construct grandly-titled pro-poor agricultural innovation systems, modes of socially-relevant and sustainable innovation have been going on all around us for many years. BRAC, Amul, Real IPM, FIPS, and Gram Mooligai are part of an expanding diversity of such initiatives, some ancient, some modern, ranging from the Chipko women's environmental protection movement in India, Systems of Rice Intensification (a low-input approach to rice production innovation), the micro-finance approach of Mohammed Yunus in Bangladesh, the Intermediate Technology philosophy of

Schumacher, to the Campaign for Real Ale in the UK and many, many more (see http://aylluinitiative.org/ for an initiative that is trying to map these globally).

To varying degrees, all of these initiatives disrupted existing modes of production and innovation. Yet many initiatives of this sort take place, because of perception and analytical reasons, out of sight and below the market and policy radar. Of course, many simply fail and only a fraction of initiatives come to national and international prominence despite the opportunities that they present for both profit and social change. Have we then misunderstood what really drives innovation and ended up overlooking major hotspots of creativity?

Having remembered this other world of pioneers and prophets, it now seem all too obvious that there are many challenges and opportunities in agriculture and rural development that conventional market-led innovation will not address, but are, nevertheless, important for social and environmental reasons.

At this point, it is worth going right back to some of the early writing on innovation in the 1930s by Austrian economist and political scientist Joseph Schumpeter. He pinpointed entrepreneurship as the key creative force, talking in terms of a continuous process of disruption, doing something different for gain and creative destruction of the old to bring in the new (1934, 1950).

According to Schumpeter, entrepreneurs were the forces required to drive economic progress, without which economies would become static and decayed. Entrepreneurs were entities that identified a commercial opportunity and organised a venture to take it forward. Successful entrepreneurship, he argues, sets off a chain reaction, encouraging other entrepreneurs to propagate the innovation to the point of 'creative destruction'– 'a state at which the new venture and all its related ventures effectively render existing products, services, and business models obsolete'. Thus Schumpeter saw the entrepreneur as operating within a larger system, sparking change by example and thriving on that change as fuel for more action.

Entrepreneurs – social, market and environmental – look for opportunities and often complement or substitute for missing action by the public sector or other actors (Mair, 2008). While attention to profits is often a consideration while still keeping in mind various developmental mandates, entrepreneurship is often not just about making money; rather, it is about adding value – social, market or environmental – in a given societal context (Dees, 1998).

Other than taking risks what does entrepreneurship involve? It means being responsible for the marshalling of ideas, resources, people, processes, institutions and policy to effect new outcomes. Of course, these are precisely the sort of processes that innovation systems ideas have been advocating, but struggling in their misguided attempt to orchestrate.

4.4 What lies below the radar?

Having recognised that it is entrepreneurship, generally, rather than the private sector or companies, specifically, that we are interested in, this then allows us to see different forms of entrepreneurship and innovation and the potential for synergies between them: market entrepreneurship (for profit), social entrepreneurship (for social change), environmental entrepreneurship (for environmental change and protection).

As the social entrepreneurship literature (Mair, 2008; Gries and Naudé, 2011) is keen to point out, these distinctions hide the fact that social enterprises often also encompass market and environmental agendas. They can also include researchers acting as social entrepreneurs because they have seen a way of making their science count. What is interesting, however, is that there is a growing class of entrepreneurs that is pioneering initiatives that explicitly occupy the intersect between market, social and environmental concerns (× marks the spot in the diagram, Figure 4.1). Those mentioned in the introduction fall into this category.

This sort of hybrid, below-the-radar entrepreneurship should not be confused with corporate social responsibility. Rather, it includes organizations explicitly aiming to be for-profit and for-a-difference simultaneously. What is critical here is that these entrepreneurs are not necessarily profit-maximizers, but individuals and groups happy to live with the trade-offs between profit and making a difference. They range from seeming social activists to disruptive innovators. They often have iconoclastic tendencies, challenging what they see as the foolish ways of old.

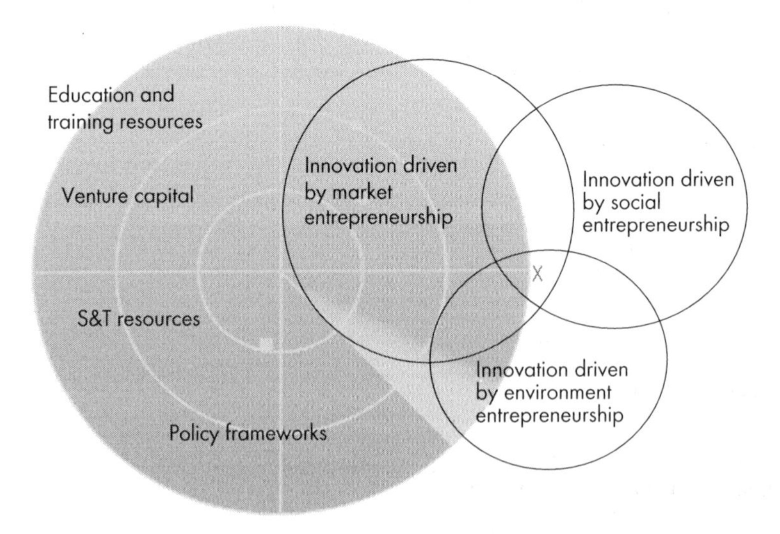

Figure 4.1. What lies below the radar?

The growing visibility of these below-the-radar entrepreneurs is driven by:

- Reaching the limits of the effectiveness of the corporate sector's capability to innovate in response to large markets of poor people (and the opportunities that this opens up for alternatives).
- Limits to existing modes of innovation in environmentally-fragile production zones.
- A global shift in the centre of innovation gravity from the Northern corporate world to new areas of economic and social dynamism in the South.

Kaplinsky and colleagues at the Open University in the UK argue (Kaplinsky *et al.*, 2010) that these new disruptive modes of innovation are going to assume prominence in India and China and that this may have global implications.

There is no standard prototype of an entrepreneur, but we are now beginning to get a fairly good idea of the features that characterize one:

- *Self-organising, opportunity-driven.* Individuals and groups see an opportunity for making a difference. This may be a market opportunity (for example, for ethically-produced goods, but also increasingly viewing the poor as a large market), a technical opportunity (for example, technological advances in renewable energy, bio-pesticides) or an opportunity arising from organizing activities in a different way for social purposes (for example, microfinance).
- *Demand regime changes.* Almost by definition below-the-radar approaches demand changes in the way policies and institutions organize the world. The Real IPM Company needed to push for regulatory changes to allow the approval and sale of bio-control agents as 'pesticides'. The successes of the organic food movement around the world happened because they were able to amend food labelling regulations. The systems of rice intensification movement continues to challenge the scientific understanding of rice production.
- *Often get stuck.* Just like in the business world, many social and below-the-radar entrepreneurship initiatives fail. Many just aren't very good. But many promising ones fail because they get stuck. Sometimes networks are not wide enough to bring in new technical support needed to cope with market or regulatory changes, as in the case of some ethical-trade initiatives. The bio-fuels sector in Senegal got stuck because of the absence of a coalition of companies and NGOs working in the sector to lobby coherently for supportive policy regime changes. The bio-gas sector in Switzerland succeeded because it did. The Intermediate Technology movement lost momentum because it couldn't move beyond engineering adaptation to deal with developing new policy frameworks. Modes of financing available are almost always too short-term to incubate initiatives that need to address both social and institutional changes – which may take 10 years or more to come to fruition.

4.5 Conclusions and implications for policy

The way forward is to start to pay much more policy attention to the support of social and below-the-radar entrepreneurs. We should stop attempting to construct new socially relevant innovation systems. Instead we should use innovation systems principles to support these hybrid, below-the-radar modes of innovation that are pioneering business models that address the most critical goals of development in most countries, not only in the South but in the North too.

There are also opportunities here for companies and the corporate sector. Many of these below-the-radar entrepreneurs are pioneering business models that are finding ways of accessing large markets of poor people – a task, which for reason of size, structure and perspective, many large companies have not been able to achieve. These models are likely to be very attractive to companies that have reached the limits of market share in existing markets.

This implies some broad shifts in policy in support of innovation:
- *From mainstream to every stream.* Supporting a diversity of entrepreneurs and modes of entrepreneurship occupying a multitude of niches that address different market, social and sustainability objectives in different ways.
- *From replicable models to a mosaic of niche successes.* Instead of looking for pilots that can be copied and transferred with an aim of new 'industry standard' approaches, the emphasis should be on identifying a wide diversity of promising enterprises and supporting their successful development.
- *Orchestration to facilitation.* Innovation is largely self-organizing (at least at first) and is driven by entrepreneurship. Policy, therefore, needs to play a supporting role. Activities such as research need to organized as a resource that can easily be drawn upon. Incubation and the creation of spaces for early stage entrepreneurial activity will be much more important.

4.6 Practical Implications

The central message for planning interventions is that the emphasis needs to shift away from attempting to construct new systems and instead focus on supporting emerging nodes of creativity. This support needs to extend to creating the conditions for below-the-radar innovation to emerge in the first place. Figure 4.2 (from World Bank, 2006) contrasts the way policy interventions can be organised in orchestrated innovation trajectories and the opportunity-driven trajectory of the sort driven by entrepreneurs. The following broad categories of intervention and support will be important.
- *Scanning.* There will be a strong analytical role for research to look below the radar for promising new innovation processes with strong social and sustainability relevance. Filtering out initiatives that are likely to fail will be a large part of this task.

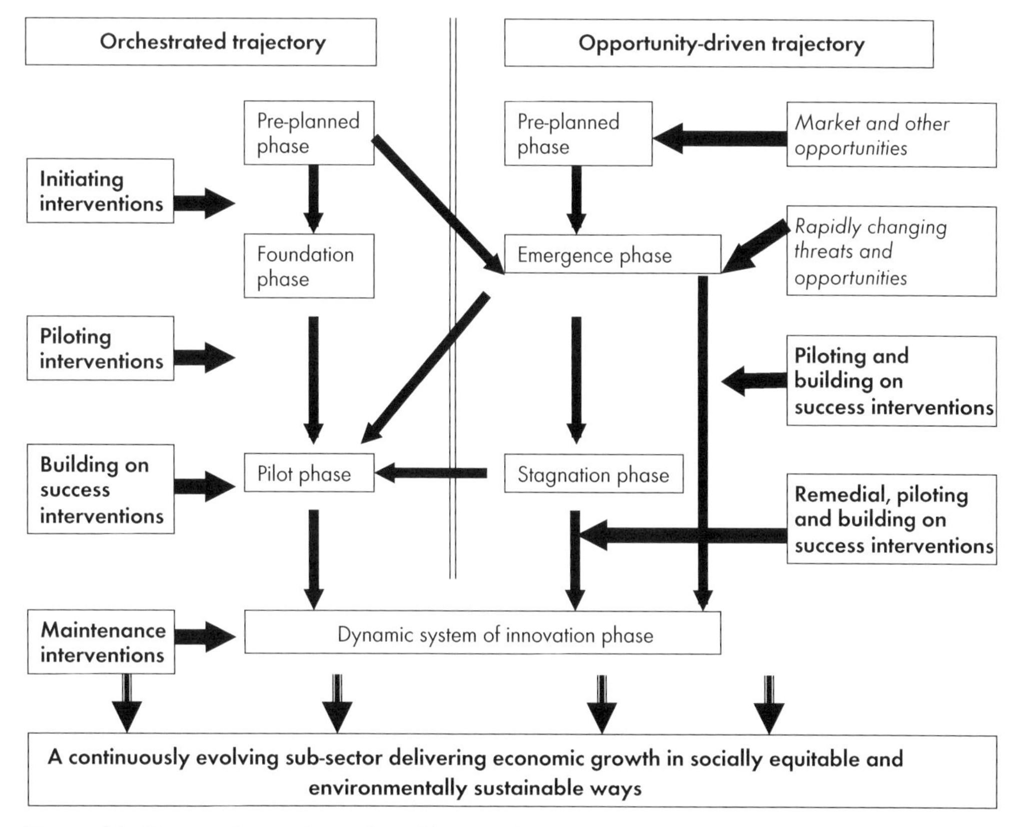

Figure 4.2. Contrasting trajectories of innovation.

- *Financing.* New forms of financing will need to be made available – not only challenge funds, but also novel types of social venture capital. It will also need to be recognized that returns to investments in these initiatives may only take place over the long-run, due to extended incubation requirements and the need for changes in policy regimes.
- *Adaptive support services.* The initiating phase of these initiatives is self-organizing, but this is also when support is needed. Adaptive services – which can not only facilitate change in technical, institutional, policy arenas but can also help with coaching, incubation training, networking, financing – are required. This will include research support.
- *Risk reduction.* Mechanisms to reduce the risks to new entrepreneurial activity – tax incentives, grants and new financing mechanisms – are important.

The ultimate paradox of all is that while we have been musing about the shape of agricultural innovation systems of the future, they have been with us all along – if only we had been able to see them. The challenge now is to find ways of structuring support for ever-shifting patterns of innovation in which the organizational categories of the 20th century are rapidly becoming irrelevant.

References

Biggs, S.D., 1990. A multiple source of innovation model of agricultural research and technology promotion. World Development, 18 (11), 1481-1499.

Biggs, S.D. and Clay, E.J., 1981. Sources of innovations in agricultural technology. World Development, 9, 321-336.

Dees, J.G., 2001. The meaning of social entrepreneurship. center for the advancement of social entrepreneurship (CASE), May 30, 2001. Available at: http://www.caseatduke.org/documents/dees_sedef.pdf.

Engel, P.G.H., 1995. Facilitating innovation: an action-oriented approach and participatory methodology to improve innovative social practice in agriculture. PhD Thesis, Wageningen University, Wageningen, the Netherlands.

Freeman, C., 1995. The 'national system of innovation' in historical perspective. Cambridge Journal of Economics, 19 (1), 5-24.

Gries, T. and Naudé, W.A., 2011. Entrepreneurship and human development: a capability approach. Journal of Public Economics, 95(3), 216-224.

Hall, A.J., 2009. Challenges to strengthening agricultural innovation systems: where do we go from here? In: Scoones, I., Chambers, R. and Thompsons, J. (eds.) Farmer first revisited: farmer-led innovation for agricultural research and development. Practical Action, Rugby, UK, pp. 30-38.

Hall, A., 2006. Public-private sector partnerships in a system of agricultural innovation: concepts and challenges. International Journal of Technology Management and Sustainable Development, 5 (1).

Hall, A.J., 2005. Capacity development for agricultural biotechnology in developing countries: an innovation systems view of what it is and how to develop it. Journal of International Development, 19(5), 611-630.

Hall, A.J., Rasheed Sulaiman, V., Clark, N.G., Sivamohan, M.V.K. and Yoganand, B., 2002. Public-private sector interaction in the indian agricultural research system: an innovation systems perspective on institutional reform. In: Byerlee, D. and R.G. Echeverria (eds.) Agricultural research policy in an era of privatization: experiences from the developing world, CABI, Wallingford, UK

Hall, A., Sivamohan, M.V.K., Clark, N., Taylor, S. and Bockett, G., 1998. Institutional developments in Indian agricultural R and D systems: emerging patterns of public and private sector activities. Science, Technology and Development, 16(3), 51-76.

Juma, C., 2010. The new harvest: agricultural innovation in Africa. Oxford University Press, Oxford, UK.

Kaplinsky, R., Chataway, J., Clark, N., Hanlin, R., Kale, D., Muraguri, L., Papaioannou, T., Robbins, P. and Wamae, W., 2010. Below the radar: what does innovation in emerging economies have to offer other low income economies? International Journal of Technology Management and Sustainable Development, 8(3), 177-197.

Leeuwis, C. and Pyburn, R. (eds.), 2002. Wheelbarrows full of frogs: social learning in rural resource management. Koninklijke Van Gorcum, Assen, the Netherlands.

Mair, J., 2008. Social entrepreneurship: taking stock and looking ahead. Paper presented at the 2008 World Entrepreneurship Forum.

Roling, B., 1992. The emergence of knowledge systems thinking: a changing perception of relationships among innovation, knowledge process and configuration. Knowledge and Policy, 5(1), 42-64.

Schumpeter, J.A., 1934. The theory of economic development. Harvard University, Cambridge, MA, USA.

Schumpeter, J.A., 1950. Capitalism, socialism and democracy. Harper & Row, New York, NY, USA.

World Bank, 2006. Enhancing agricultural innovation: how to go beyond the strengthening of research systems. Economic Sector Work report. The World Bank, Washington, DC, USA.

Part II. Addressing new issues

Chapter 5. Innovating in cropping and farming systems

Jean-Marc Meynard

5.1 Introduction

Intensive farming systems, those that are heavy users of chemical inputs (fertilizers, pesticides, veterinary products) and are largely mechanized, today dominate agriculture in developed countries and are expanding in many countries of the South. They are generally more productive per hectare than traditional agricultural systems and are labour-efficient, leading to an unprecedented rise in labour productivity. However, the question of the sustainability of these input-intensive systems must be considered. They are heavy fossil-energy consumers, producers of greenhouse gases and are generally detrimental to biodiversity. They pollute water with nitrates, phosphoric compounds and pesticides. While they may have helped reduce famine and lowered urban food costs, they have also led to the concentration of production in the hands of those who have been able to invest in inputs and mechanization, and thus have exacerbated the vulnerability of farmers who had no access to these means (De Schutter, 2010).

The questions now arise: should these input-intensive systems still be promoted, albeit after correcting their most serious drawbacks, or should they be replaced by other forms of agriculture? What production systems can we imagine to ensure food security for rural and urban populations which will generate regular and sufficient remuneration to farmers, while, at the same time, conserving natural resources and promoting the adoption of ecosystemic services and social cohesion at the territorial level?

In this chapter, we analyze the agronomic rationales of current systems – and their economic and social determinants – using various examples from France and countries of the South. In doing so, we assess the possibilities of changing them and the steps that would be required to render them sustainable. We show that the agronomic rationale will have to be changed and agricultural systems reinvented by calling upon agroecological engineering approaches. We draw some proposals concerning the orientation of agricultural R&D and of actions undertaken by authorities in order to promote the necessary changes.

5.2 Agronomic, economic and social rationales for input-intensive agricultural systems: an example from France

Intensification of input use is part of an overall rationale that must be analyzed simultaneously at the agronomic, economic and social levels, encompassing not only farms but also territories and supply chains. We shall illustrate the interconnection of these various levels through the case of cereal farming systems in the Seine watershed, which is a major part of the Paris basin, in northern France.

5.2.1 Specialization of territories and production systems

Since the 1960s, when they dominated agriculture in the Paris basin, mixed farming systems have been in steady decline. Regions in the centre of the watershed, where the land is most fertile, have specialized in cereal farming with animal husbandry almost disappearing. In peripheral areas, cattle farming was intensified based on the development of silage maize and on soya bean meal import. Figure 5.1, taken from Schott *et al.* (2011), illustrates the consequences of this change in terms of land use: the maps show the evolution in cropping patterns in the Seine watershed (the Normandy coast to the north-west, Bourgogne to the east), an area of around 100,000 km^2, in censuses conducted by the French Ministry of Agriculture between 1970 and 2000. We observe an increase in cereal surface areas, mainly in wheat, especially so in the central areas of the watershed (Ile-de-France, Beauce, Brie, Picardy) where wheat covered more than 50% of agricultural surface areas of certain regions in 2000. The reduction in ruminant livestock farming in this central area is reflected by the sharp decline in natural grassland surfaces (not shown), which, in 2000, only have a significant presence in peripheral areas, such as in Normandy to the west and Thiérache to the north. Specialization towards intensive livestock farming in these regions is marked by a spectacular increase in coverage of silage maize (Figure 5.1b). This specialization of farms and entire regions poses many ecological problems: low levels of recycling of mineral nutrients (N, P, K, etc.) on farms, leading to a waste of non-renewable resources and to pollution of water (nitrate, phosphorus) and air (ammonia, nitrous oxide); loss of biodiversity due to the replacement of grasslands by annual crops; reduction of the diversity of habitat mosaics, detrimental to biodiversity; etc.

The case of alfalfa, shown in Figure 2 (Schott *et al.*, 2011), is emblematic of this process of regional specialization. In the 1970s, alfalfa was present throughout the Seine watershed, usually consumed by livestock on the farms where it was grown. Its decline followed that of mixed farming systems and, between 1980 and 2000, alfalfa became concentrated in the chalky soils of the Champagne region, an area conducive to high production (deep soils with very high available water capacity), to supply the 'dehydrated alfalfa' supply chain. Alfalfa is dehydrated using fossil fuels so that it can be easily stored, transported and incorporated into animal feed sold to farmers in areas specializing in animal husbandry. The 2000s marked a decline in this energy-consuming sector due to high energy prices, reduced European support and competition from soya bean meal imported from the Americas.

This specialization of territories is driven by the agro-industrial establishments which ensure product outlets: the dairy industry, which lifts the bulk of milk production, is concentrated in areas of livestock farming; and, conversely, central regions have been structured for collecting

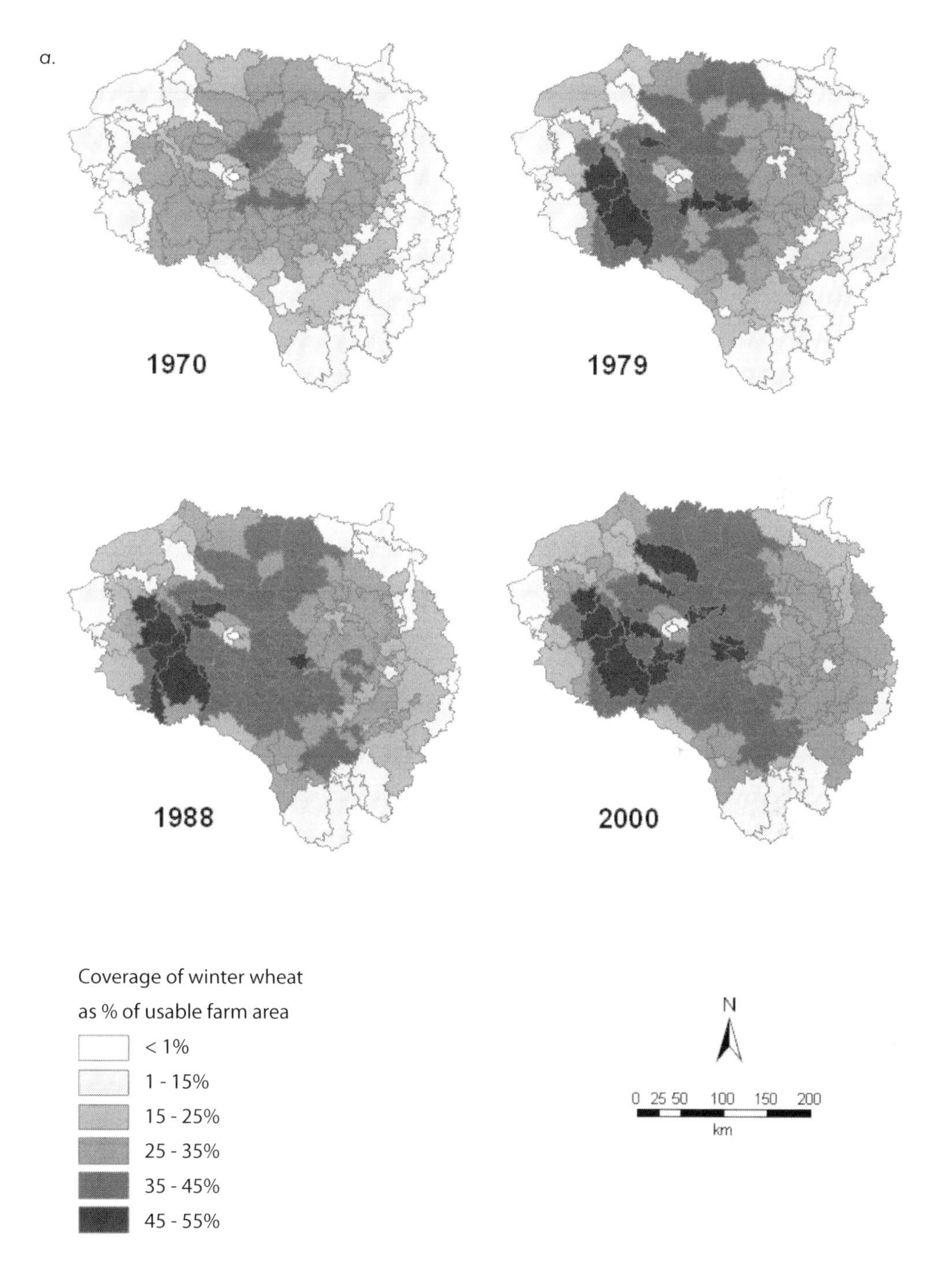

Figure 5.1. The specialization of the territories in the Paris basin: cereal farming in the centre, livestock farming at the periphery. Changes in (a) common wheat coverage and (b) silage maize coverage between 1970 and 2000 (Schott et al., 2011).

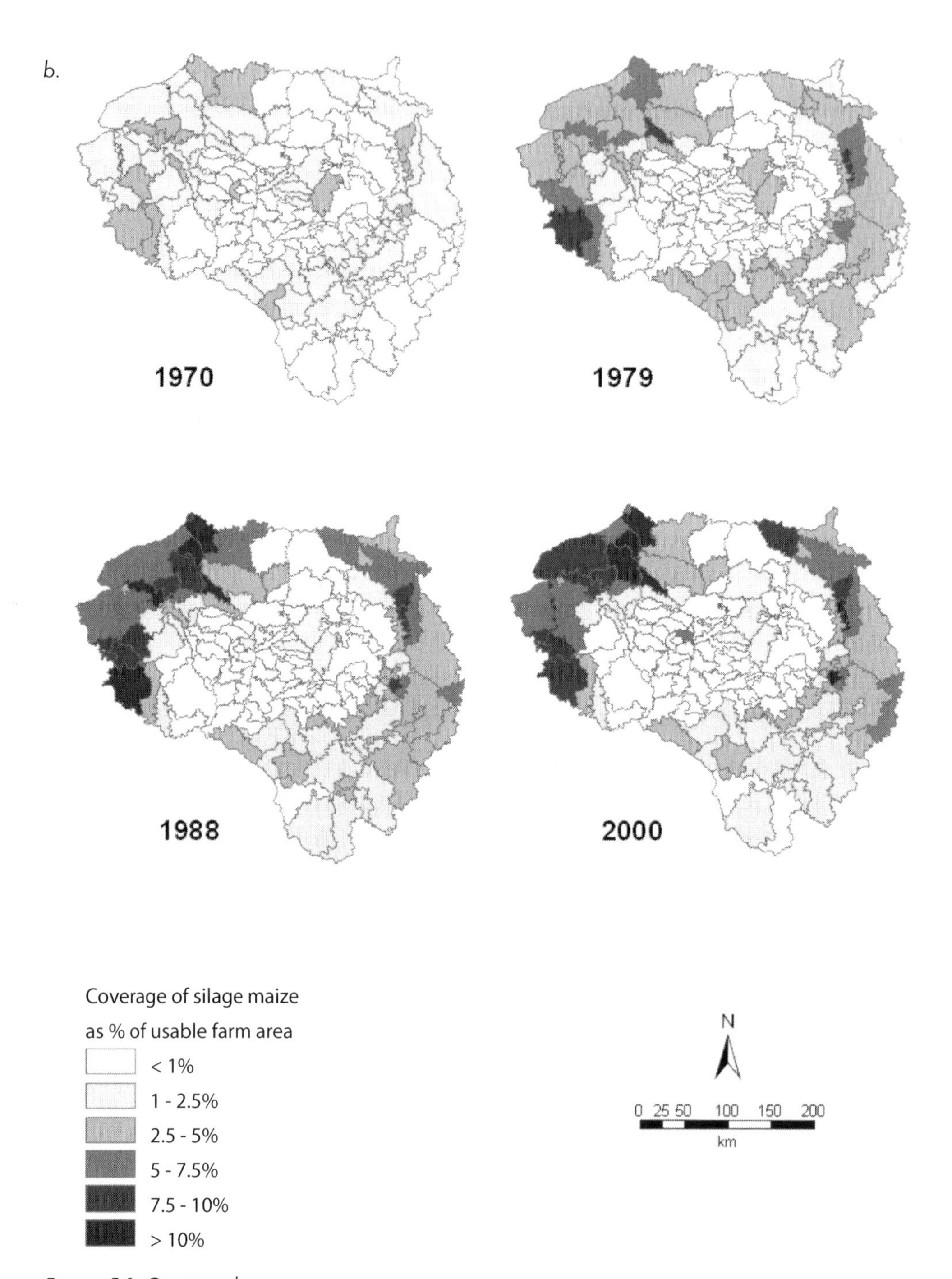

Figure 5.1. Continued.

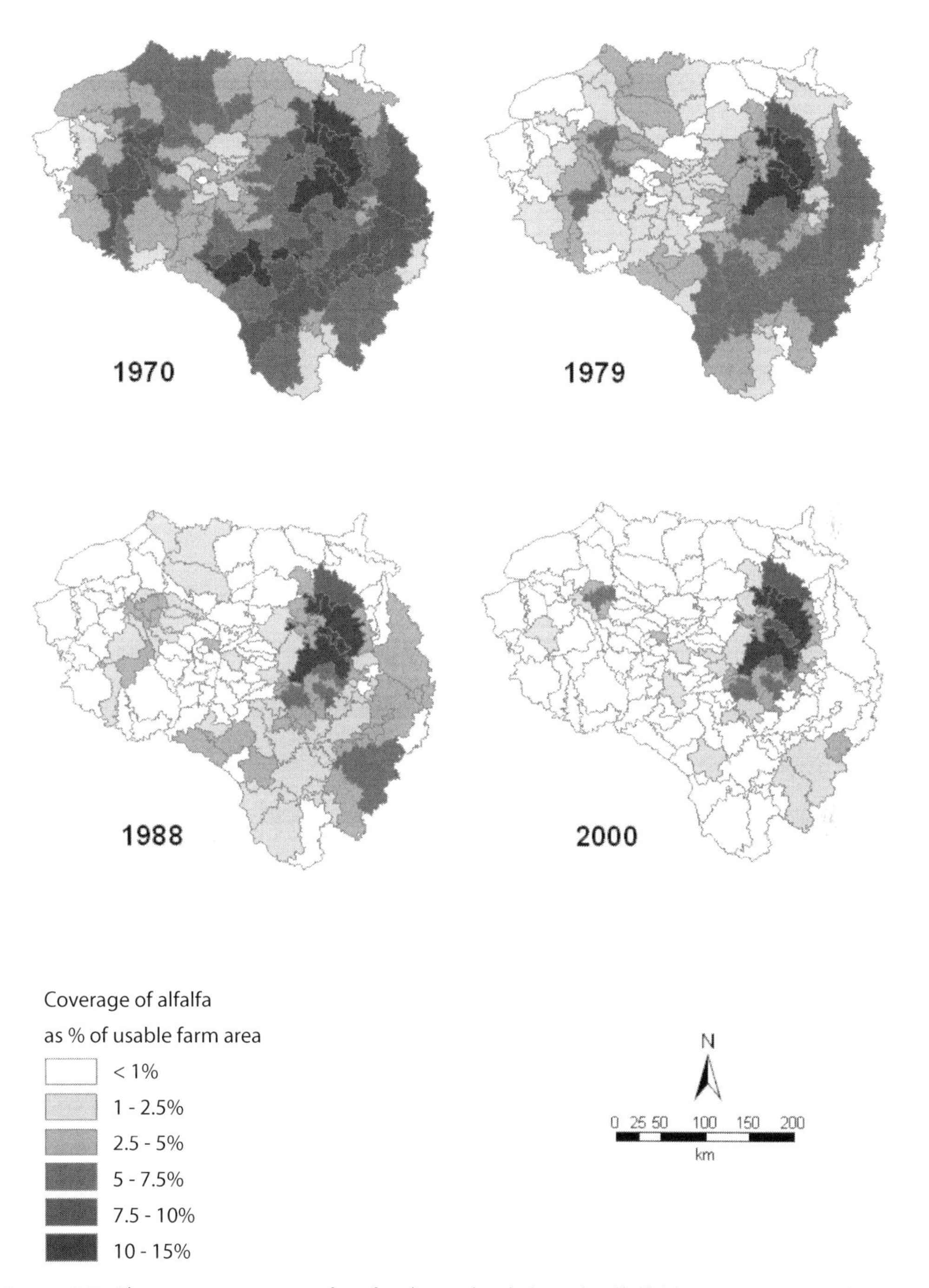

Figure 5.2. Changes in coverage of artificial grasslands (mainly alfalfa) between 1970 and 2000 in the territories of the Paris basin (Schott et al., 2011).

cereal products and sometimes processing them (sugar refining, starch production, for example). The specialization is accompanied by an increase in the technical sophistication of production methods with specialized farmers, as also their advisors, concentrating on acquiring new skills only in their specialized fields. Re-diversifying becomes increasingly difficult for a farmer. He must not only find outlets for his produce that no longer exist locally but must also acquire knowhow and technical references for new products without local support.

5.2.2 Reduction in the number of cultivated species

This regional specialization is accompanied by a reduction in the number of species cultivated and shorter rotations. Thus, in areas of cereal farming in the Seine watershed, the surface areas of wheat and rapeseed have increased significantly between the 1980s and 2000s. In contrast, the surface areas of protein peas, sunflower and maize have diminished while wheat monocultures, or short rotations, such as rapeseed-wheat-wheat or rapeseed-wheat-barley have increased in frequency (Schott *et al.*, 2011). The entire sector is involved in this simplification of cropping plans: the example of the decline of the pea crop is illustrative (areas reduced to 1/7th in 15 years in France, between 1994 and 2009). The increase in coverage of pea crop in the 1980s was due to sustained government support (guaranteed high prices, support for varietal selection and the development of technical references), linked to a desire of the European Union to reduce the dependence on protein resources for animal feed. The decline in area under cultivation started with a reduction in governmental support but was also related to the spread of a new telluric disease: *Aphanomyces euteiches.* During the period of maximal expansion of pea surface areas, at the turn of the 1980s and in the 1990s, some farmers had indeed grown peas on the same fields a little too often, which favoured the spread of the disease. A decline in surface areas discredited the pea in the eyes of cattle feed manufacturers, who could no longer count on a regular and guaranteed supply. Consequently, soya bean meal became the essential source of protein for cattle feed, and pea prices suffered, further reducing areas under cultivation. Private seed companies became sceptical about the pea's future and reduced investments in breeding new varieties. The productivity differential with wheat or rapeseed – which are seeing no reduction in efforts of varietal selection – is not diminishing, a prerequisite to rekindle economic interest in protein pea cultivation. It is indeed the combination of the various strategies of the actors in the sector that have led to the decline in pea cultivation and which will singularly complicate the reintroduction of peas in cropping plans, as desired now by the French government.

5.2.3 The key role of pesticides

Shorter rotations increase the problems of soil parasitism and make it difficult to control weed populations. The homogenization of cropping patterns increases the risk of growth of air-borne parasite populations (insects, foliar fungal diseases). The process of specialization that we have described above would not have been possible without the use of pesticides. As an illustrative example, we can observe that regions with the greatest amount of rapeseed cultivation in the Seine watershed are also those in which the per-hectare use of pesticides for rapeseed is the highest (Schott *et al.*, 2011). In fact, pesticides have become the cornerstone of current intensive cropping systems and dictate not only the rotations but also planting dates and varietal choices. The example of winter wheat management is significant in this respect (Lamine *et al.*, 2010; Meynard and Girardin, 1991). To maximize productivity, wheat is sown early and at high density. High levels of nitrogen fertilizer are applied and varieties are chosen based on productivity, not on their resistance to disease. All these choices are favourable to high levels of production but they also increase the risk of parasitic insects, fungal diseases or weeds. Faced with these risks, a heavy application of pesticides on the crops becomes imperative. To facilitate rapid and large-scale interventions, the farmers acquired big and powerful equipment and sought to consolidate and enlarge their fields. This is how the landscape came to be dominated by cereals, through intensive farming on large fields which are genetically homogeneous and require heavy pesticide use.

This key role played by pesticides in cereal farming regions is reinforced by the advisory systems and crop breeding priorities. Given the importance of pesticides in the rationale of cropping systems, companies that market these inputs have become the main source of advice to farmers. For fighting bio-aggressors, this advice is tilted towards the use of chemical solutions, simple and dramatically effective (one problem, one solution), rather than towards preventive agronomic methods, more complex to implement and less directly effective (Butault *et al.*, 2010). Resistant varieties are most often considered as adjuncts to pesticides, not as the primary means of pest control. As a result, the market for multi-resistant varieties remains limited, with plant breeders unwilling to concentrate on what remains a niche market. Given the secondary role of resistant varieties, there is no coordination of varietal choices for a durable management of resistance. Varietal resistance is thus regularly circumvented which tends to discredit the genetic solution.

5.2.4 A technological lock-in

Production systems in the area studied are thus fully in harmony with upstream and downstream supply chains, as well as with information dissemination systems. Each actor's strategy strengthens and reinforces those of the others. No one has any incentive to change strategy or direction as long as the others also do not change. This is a typical case of 'technological lock-in[2]' into specialized and input-intensive agricultural systems. Effects of lock-in into pesticide use have already been described in different countries: Cowan and Gunby (1996) in the USA, Vanloqueren and Baret (2008) in Belgium and Lamine *et al.* (2010) in France. The overriding conformity of socio-technical systems which produces this lock-in

[2] The concept of technological lock-in reflects a situation in which a technology A could be adopted permanently, even irreversibly, at the expense of technology B, even if technology B appears, *ex post*, to be the most effective (Labarthe, 2010).

is the result of the remarkable response of the agricultural world to the demand for increased cereal production and for increased competitiveness of agriculture. This conformity clearly clashes with any new orientations that environmental considerations may suggest. As emphasized by the work of economists on the abolition of milk quotas (Daniel *et al.*, 2008) or on crop diversification (Farès *et al.*, 2012), the economic mechanisms at the supply chain level tend to reinforce this lock-in. According to the concept of 'path dependency' described by economists of innovation (David, 1985; Dosi, 1988), innovations which are consistent with this lock-in (ever shorter rotations, tools to help decide on amounts of inputs to use) have a greater chance of being adopted than those that are not (Labarthe, 2010). The changes allowable in systems are strictly limited; opening wide the possibilities of innovation-led changes is not possible without 'unlocking'.

5.3 What leeway for changing input-intensive systems?

Socio-technical systems similar to the one described for the Paris Basin in France predominate in developed countries. But they are also spreading in agricultural regions of countries of the South. To understand how far this is a worldwide phenomenon of the evolution of agricultural systems, rooted in decennial dynamics and based on international actors and rules of world trade, we will take up three emblematic examples of this evolution, in contrasting ecological and socio-economic contexts: the spread of soya bean in the Argentinean Pampas, cotton production in Thailand and banana production in the French West Indies.

5.3.1 The spread of soya bean ('sojizacion') in the Argentinean Pampas (adapted from Grosso, 2011)

Since the mid-1990s, there has been a rapid expansion of cereal farming, especially soya bean, in the Argentinean Pampas at the expense of grasslands and animal husbandry: the area under soya bean coverage has grown four-fold (from 500,000 to 2,000,000 ha) between the early 1990s to the end of the 2000 decade. Soya bean varieties that are grown are, almost all, herbicide tolerant (glyphosate in particular). This expansion of soya bean is directly related to the fact that its gross margin is greater that those of other crops. About 60% of cultivated land is currently planted with soya bean and increasingly rotations include soya bean each year. The growth in soya bean has been accompanied by an almost universal elimination of tillage, thus reducing working time and mechanization costs. There exists a clear synergy between the two innovations: direct seeding and herbicide-tolerant varieties. The elimination of tillage requires very effective herbicides, which is most easily obtained by the use of total herbicides associated with tolerant varieties (soya bean but also maize) rather than with conventional weed control strategies. Since the early 2000s, the growth in economic margins of cereal crops (and, in particular, of soya bean) has been much greater than that of the cost of living, thus strengthening the soya bean sector, accelerating the decline of the livestock sector, attracting investors and promoting land consolidation. We are thus witnessing the rise of financial agriculture, based on a pursuit of short-term profits from several hundred thousands of hectares of cropland, with little interest in managing rotations and long-term soil fertility ('pool de siembra'). With the increase of soya bean coverage and shortening of rotations, we see the development of viral and fungal diseases, and the emergence of weed resistant to glyphosate, resulting in increased pesticide use. The growth of the soya bean sector, mainly export oriented, has improved the economic performance of Argentinean agriculture but is being criticized for the associated environmental pollution (Botta *et al.*, 2011) and negative social impact, in particular the breakdown of local territorial dynamics (Albaladéjo, 2012).

5.3.2 Cotton production in Thailand (adapted from Castella *et al.*, 1999)

In Thailand, the surface area planted with cotton has been following a cyclical pattern since the 1960s: a gradual increase in coverage and, consequently, in production, dependent on the intensification of inputs, followed by a sharp fall in yields, leading to a reduction in coverage. A new cycle then starts, 10 to 15 years after the previous one. Intensification takes the form mainly of the adoption of 'modern' varieties, with very high yields but with less tolerance to insects than traditional varieties, and the use of the newest insecticides. Over time, with the increase of coverage and applications, a strong selection pressure is exerted on parasitic cotton insects with some of them becoming resistant to insecticides. After a phase of stronger doses and increased number of applications – a strategy that only increases the selection of resistant sub-populations – the insecticides lose all their effectiveness, the insect populations become uncontrollable and yields plummet. The cotton crop is then replaced by other crops until a new family of insecticides appears and is distributed. Castella *et al.* (1999) show that all the actors connected with agriculture contribute, in one way or another, to the success of the 'intensive cotton' technical package: agricultural development services, governments (who wish to increase cotton production) and agro-chemical companies. Local traders, who are in close relationship with many farmers (they sell them seeds and pesticides, and buy their products), use their dominant position to impose the technical package on indebted farmers. A transition to integrated pest management (IPM) in order to break this vicious cycle would require, according to Castella *et al.* (1999), the coordinated mobilization of a wide range of actors: the authorities, seed breeders, farmers, traders, researchers, advisors, etc. Such an approach is slow in developing and, since the early 2000s, cotton coverage is steadily declining and is, today, at a very low level: less than 10,000 ha, compared to 152,000 at its peak, in 1981.

5.3.3 Banana production in the French West Indies (adapted from Clermont-Dauphin *et al.*, 2003)

Intensive monoculture of banana has been practised on the volcanic soils in the highlands of Guadeloupe (French West Indies) since the 1980s. Every three to five years, banana trees are replanted in the form of nematode-free vitroplants. The organic residues from the previous banana plantation are buried by ploughing. Massive doses of fertilizers are applied and several herbicide, fungicide, insecticide and nematicide applications follow. A study of soil biological activity in sample fields (some less intensified, deliberately over-represented) shows that ploughing promotes an increase in nematode populations and instead reduces those of earthworms (which, according to Lavelle *et al.* (2004), have a depressive effect on nematode populations). Nematicides used to overcome the nematode populations appear more effective against earthworms and less harmful nematode species than against the most harmful species. Because of this vicious circle, a few years after planting, nematode populations become too high and the soil is ploughed anew to start a new plantation. We find similar production systems in all the main regions where banana is grown for export. As in the other examples, the upstream and downstream parts of the supply chain are in full coherence with this standardized production, favourable to the satisfaction of mass markets. However, in the French West Indies, this production system finds itself totally destabilized by the opening of the European market to bananas grown in countries with cheap labour and by the realization of the amount of environmental damage caused by the use of pesticides over the last few decades (Blazy, 2011). The French banana sector is therefore seeking a change in its practices so that its specificities can be recognized on the markets. In the framework of the Sustainable Banana Plan, the Institute for Tropical Technology, CIRAD and producers are engaged in an innovative approach to reduce input use (Dorel *et al.*, 2011; Blazy, 2011).

5.3.4 What leeway to change locked-in agricultural systems?

All these systems with high levels of input use share a similar evolutionary process: driven by the need for maximizing short-term profits, they are all characterized by regional specialization, shorter rotations (even monocultures) and a key role assigned to pesticides. They also have quite similar environmental and social impacts with biodiversity and associated ecosystems in decline everywhere, as much from the homogenization of landscape mosaics as from uncontrolled pesticide use. Territorial specialization does not promote the recycling of fertilizing nutrients, and these systems result in a waste of resources. Socially, the impact of pesticide use on farmers' health has become part of policymakers' agenda not only in metropolitan France but also in the French West Indies and Asia. The reduction in agricultural employment resulting from input intensification and mechanization are emptying rural areas in Argentina, as they have already done in France. Finally, the agricultural systems of the Paris basin and of the Argentinean Pampas are highly interconnected, so that French intensive farming bears some responsibility for the evolution of the Pampean systems and their impacts, in a similar way that soya bean from the Pampas contributes to the pollution of aquatic systems in western France.

These input-intensive systems display a significant agricultural coherence. To make them less polluting, less monotonous or more abstemious, it will not be enough to change one or two practices, their entire rationale will have to be changed. Their locked-in state is as much the result of their high levels of internal coherence as it is to the combined strategies of upstream actors, downstream actors and authorities. In some cases, the boom-and-bust cycles of the system (cotton in Thailand or banana in the French West Indies) have led to major crises, forcing radical changes in production methods. Elsewhere, current systems are securely anchored; there is no magic wand to wave and wish them away, neither can they be changed simply by efforts to educate the actors. To bring about such a change, it is essential to think and act systemically, avoiding being misled by solutions whose apparent simplicity may hide indirect effects difficult to control. Remember pesticides! Between the 1960s and 1990s, they were seen as a miracle solution, a silver bullet, and research, misled by the effectiveness of chemical control of bio-aggressors, invested little ressources in alternatives to pesticides (Vanloqueren and Baret, 2009); none of those alternatives is as effective as pesticides but when combined in IPM strategies, they offer interesting solutions. It is a matter of working as much at the level of agroecosystems as at socioeconomic ones on various complementary fronts: designing and proposing innovative cropping systems based on an understanding of agroecological processes; reorganization of training programmes and information distribution channels; public policies that offer incentives not only to farmers but also to other agricultural and territorial actors (De Schutter, 2010).

However, even if we can agree on the need for drastic changes in agricultural systems to make them true components of sustainable development, it would be presumptuous on our part to try to define what form they should take. We do not present here a list of technical systems deemed to be 'sustainable' or 'agroecological', which would suggest that there indeed exist universally applicable 'good' solutions that everyone would only need to implement. We will endeavour rather to recall the principles of an agroecological engineering approach to help stakeholders develop, individually or collectively, their own solutions. Given the diversity of situations and the uncertainties of the future, we propose to work on tools and approaches to help grassroot actors change their technical systems and adapt them to the specific characteristics of their situations. We will use the example of intensive cereal cultivation in northern France as a common thread in the next section of this chapter.

5.4 Tools and approaches for redesigning agricultural systems: some lines of work

As De Schutter (2010) emphasizes, agroecology offers an interdisciplinary framework for conducting research, in close partnership with the concerned stakeholders, on the transformation of agricultural systems. This framework, covered in several review articles (Wezel *et al.*, 2009; Francis *et al.*, 2003), mobilizes the agronomic, ecological and social sciences together with local empirical knowledge to design sustainable food systems. Agroecology is defined as the integrative study of the ecology of food systems (Francis *et al.*, 2003). These authors point out that it is 'impossible to deal effectively with the complexity of resource use and design of future systems if we only focus on the production aspects, short-term economics, and environmental impacts in the immediate vicinity of farm fields. It is logical to suggest that agroecology should deal with all actors in food systems as well as the total flow of energy and materials from their sources through production and other steps to the consumer, and the potential to return nutrients to the field'.

A redesign of agricultural systems along agroecology principles cannot ignore the existence of tensions, sometimes strong, between conflicting objectives: the traditional tension between economic and environmental requirements; tensions between territorial actors having differing objectives; and tensions between the rationale of individual actors and their consequences at the landscape level (e.g. all the farmers choosing to grow the most profitable crops or varieties, thus creating a genetically homogeneous space conducive to the spread of diseases). It will thus be a matter of arriving at compromises, at different scales, between these conflicting objectives (Meynard *et al.*, 2012). Another, equally important, issue is to facilitate the adaptation of agricultural systems to climate change, which could result in an increase in climatic hazards (raised temperatures, droughts, floods, etc.) and parasitic risks (migration of parasites from warm to temperate regions). Climate change challenges the frames of reference of agricultural actors and of agricultural and territorial development by questioning the relevance of knowledge acquired in past years. In such a context, it will therefore be useful to reinforce the actors' ability to project themselves into the future by combining the locally relevant knowledge acquired in their territories of action with knowledge derived from climate and agroecological modelling by scientists.

A redesign of agricultural systems involves various categories of actors. It is by these categories that we divide the rest of this section:
- tools and approaches designed for farmers and their direct advisors;
- tools and approaches to help territorial actors (e.g. farmers, local governments, environmental groups, resident, etc.) explore coordinated management scenarios.
- approaches to help authorities support the transformation of agricultural systems and promote cooperation between the territorial actors.

5.4.1 Tools and approaches for farmers and their technical advisors

Proposals, originating from R&D, of new technical systems for farmers are many and varied: tools for improving and fine-tuning fertilizer application, irrigation or pesticide treatments; no-till systems; resistant varieties, biological controls and prophylactic methods combined in integrated pest management systems; combination of varieties and species to reduce use of inputs and improve their efficiency; and self-sufficient feeding systems based on legume-rich multispecies pastures...

But in order for farmers and advisors to be able to adapt agricultural systems to the diversity of soils, agroecosystems and farms, they have to be offered methods and approaches for building their own solutions. Meynard notes that the approaches for designing innovative agricultural systems fall into two main categories: '*de novo*' design or 'step-by-step design' (Meynard, 2008; Meynard et al., 2012). *De novo* or clean-slate designs are those that break from the past, originating from design workshops or *in silico* explorations of innovative technical systems using a computer modelling tool (Bergez *et al.*, 2010; Rossing *et al.*, 1997). Step-by-step or incremental designs avoid breaks with the past and instead attempt a gradual transition to innovative systems, based on learning loops. These two families of approaches are complementary:

- The *de novo* design opens the field of possibilities by unleashing creativity and thus allows very innovative solutions to be considered, even those that might be incompatible with today's socio-technological systems but whose exploration helps prepare for the future.
- In the step-by-step design, exploration is more cautious but it has the advantage of being easily adaptable to the specific constraints of each agricultural situation. The farmer, often supported by a technician or a group of his peers, develops *his* new system himself, and, at the same time, lets himself be persuaded of the effectiveness of this new system.

See Box 5.1 for an example of the dynamics involved in learning to undertake a step-by-step design process for reducing input use and environmental pollution in a farm in Picardy (adapted from Mischler *et al.*, 2009). In this specific case, the issue was to include environmental criteria in the training of farmers as part of the redesign. Indeed, each year, farmers use their observations to further their experience: they observe soil and crop reactions; they scrutinize the behaviour of the new variety that they have tried; they test the innovation suggested by a technician in one corner of a field; etc. However, the criteria which they use for their assessments are, for the most part, mainly related to production (yields, quality) and to the economic performance of their farms. For them to embark on a redesign of their systems, they need the world of research to provide them with:

- A full set of diagnostic indicators, easy to measure and appropriate. Interest and activity in the development of such agroecological indicators is currently at a high level (Bockstaller *et al.*, 2008; Sadok *et al.*, 2008).
- A library of innovations (varieties, pesticides, decision-making tools, but also methods for integrated pest management, methods for organic matter management, diversification crops, examples of innovative cropping systems designed via '*de novo*' processes, etc.). For the farmer to make an informed choice, it is important for each innovation to be described and characterized accurately: working time; equipment and skills necessary; expected impact on the environment and the production; systemic effects on other practices; etc.

Farmers cannot usually undertake these changes alone since they run against established practices, knowledge, social representations and organization of work. Sociologists have shown (Darré, 1994; Lamine *et al.*, 2009; Warner, 2007) the significant role that farmer discussion groups can play in encouraging the adoption of innovative systems by being a source of ideas and a platform for the sharing of experiences. Farmers also find there moral support for the risks they are taking in embracing the new.

5.4.2 Tools and approaches for coordination at the landscape level

Actions undertaken at the level of individual farms are obviously ineffective in managing processes that have impacts at the landscape scale (erosion, effects on biodiversity, pollution of aquifers). These actions must be accompanied by coordination between neighbouring

Box 5.1. Dynamics of learning at the farm level: 'step-by-step' design of an agroecological production system in Picardy (adapted from Mischler et al., 2009).

The farm belonging to M.X., a farmer in the Picardy region of France, specializes in field crops (cereals, oil seeds and grain legume crops, sugar beets). It is being monitored under the framework of the 'Integrated Farming' programme of Agro-Transfert Ressources et Territoires, the Picardy Chambers of Agriculture and INRA.

In 2002, an agricultural and environmental diagnosis of the farm was conducted by the farmer and an advisor. A major weakness was revealed: extensive pesticide use (index of frequency of application (IFT in French) of pesticides over 8, which means that the farmer applied 8 recommended doses, on the average, on the fields in his farm), on relatively undiversified rotations. The technician and the farmer then explored all possible solutions: What new crops can be grown? For which markets? Are they compatible with the farmer's equipment? With his organization of work? What is the method to use, for each of the crops, to limit pesticide use? Is it possible to switch to mechanical weeding? What new varieties? Can one consider associations of species or of varieties? From the 'innovation library' offered by the advisor (based on his experience and knowledge originating with research), the farmer made his

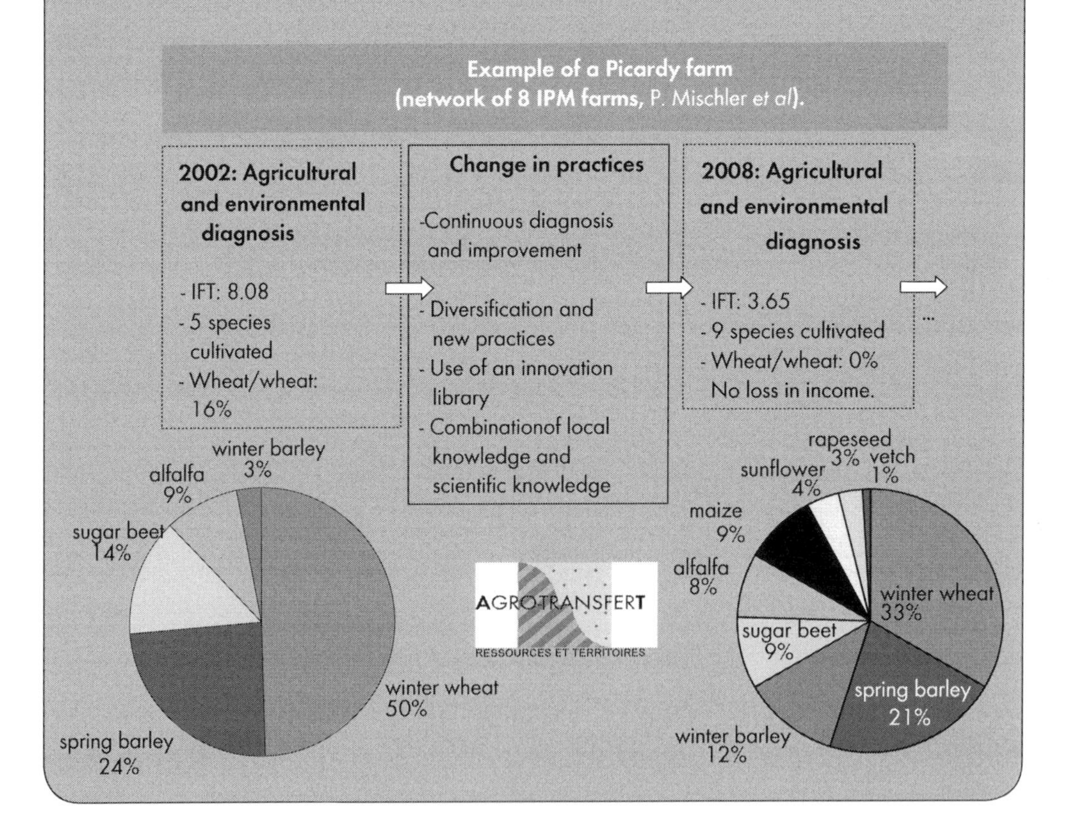

Box 5.1. Continued.

choice based on his specific constraints. He then tried out the innovations in one or two fields, then extended them to the entirety of his farm. Every year, he conducts a review with his advisor and a group of farmers who have adopted the same approach: which innovation was successful? Which failed? How to improve the initial plan of action?

Six years later, he has diversified his rotations, changed his varieties and cropping patterns, improved his energy balance (lowered use of nitrogen fertilizer and reduced tillage) and lowered the average IFT of his farm to 3 (resistant varieties, longer rotations, mechanical weeding). A follow-up evaluation conducted by the research team shows that the working time has slightly increased but income has not been affected.

farms and, more generally, between the different territorial actors to conceive landscape mosaics and spatial arrangements of agricultural systems (Papy and Torre, 2002; Soulard, 2005; Thenail *et al.*, 2009).

How to encourage such coordination? An answer is not easy to come by because the interests of the different actors may be conflicting, their perceptions of the situation irreconcilable, or the information they have asymmetric. Some may not derive – or perceive not to derive – any benefit from coordinated management of farms. Existing stakeholder networks may promote (or sometimes hinder) the innovative coordinated activities. In such a context, participatory research is particularly suitable because it is conducive to collective learning processes, which help in the construction of collective norms and coordinated practices. Among participatory approaches, Integrated Assessment (Bland, 1999; Pahl-Wostl, 2005) and Companion Modelling (Collectif ComMod, 2006) are based on the use of simulation models to mediate between stakeholders. The results of the models are used by them to compare, discuss and evolve scenarios, encouraging a convergence of stakeholder views. For example, companion modelling is based on a sequential approach (Etienne, 2010): (1) Constructing a shared representation of the territorial processes that have to be managed collectively. In conjunction with the stakeholders, researchers develop a computer model which, based on the knowledge of local actors combined with that of scientists, explains key interactions between ecological, agricultural and social systems. (2) Organizing a role playing game, which, by placing actors in situations, allows them to learn and understand, from within, the complexity of interactions. (3) Comparing scenarios constructed on the basis of the role playing game. These different scenarios result from different strategies adopted by certain actors or from different regulations, for example. The model helps stakeholders weigh the pros and cons of the different scenarios and improve them, thus approaching more satisfactory solutions.

Advances in research on the modelling of spatial processes can be a valuable resource for testing, jointly with the territorial stakeholders, various scenarios of land-use and of changes in agricultural systems. Agronomists and ecologists must come together to model interactions between technical systems and agroecosystems for this purpose. The methodological impediments to such an interdisciplinary task are many: complexity of translating technical actions into ecological variables; differing views on the temporality of processes; disagreements on the suitable subdivision of space (the landscape cannot be reduced to mere fields that are so dear to the agronomist, nor to a single encompassing 'landscape structures' that the landscape ecologist may feel comfortable with).

5.4.3 Some thoughts on public action

The impact of government action on the evolution of agricultural systems and landscapes is potentially huge. Taking again the example of the Seine watershed, mentioned at the beginning of this chapter, in the past, the support price of wheat contributed to the disappearance of livestock farming wherever areas were favourable for cereal farming; drainage incentives resulted in the regression of natural wet meadows and associated ecosystemic services (biodiversity, water filtration, etc.); irrigation incentives, created by the Agricultural Policy of the 1990s, have led to a proliferation of irrigation users and maize monocultures, with sometimes serious consequences for the ecology of aquatic environments (Meynard *et al.*, 2003). Crop specializations, shortening of rotations and the development of intensive cropping systems that consume large amounts of pesticides were thus all encouraged by the rules laid down by the government.

The economic instruments that can be used to manage agroecosystems are well known: public aid (agri-environment measures, conditionality, support for the dissemination of innovations); taxes (on pesticides, on energy, etc.), obligations or prohibitions; and quotas. These tools could be used better to pay, directly or indirectly, for the ecosystemic services provided by agriculture. However, an insufficiently systemic vision sometimes leads to perverse effects: for example, the requirement to cover land in winter by planting a catch-crop (to reduce the risk of nitrate pollution of water) promotes, in some environments, the multiplication of slugs, and prevents farmers from stubble breaking and stale seedbed, techniques which lead to a reduction in herbicide use. In the same way, in Europe, the withdrawal of milk quotas announced for 2015 will inevitably lead to the concentration of production in the hands of the most competitive operators and of milk collection in regions with high producer densities. The specialization of farms and territories is consequently bound to increase.

Any effort on the part of governments to promote crop diversification through regulatory means or financial support will work over the long term only if it is backed up and perpetuated through market mechanisms. The diversification of landscape mosaics, as favourable to biodiversity as a reduction in pesticide use, requires government support for the creation of new supply chains and sectors: it should help diversification sectors to emerge, consolidate and gain credibility. It would require, among others, coordinated action: (1) for the selection of 'orphan' species (role of government research, targeted support to private selection efforts); (2) at the agricultural R&D level (targeted support for the development of references for diversification crops); and (3) at the downstream supply-chain level to promote technological innovations and coordination between actors.

On the other hand, we have seen the major role that dynamics of local and collective learning can play in transforming agriculture. What can governments do to promote these dynamics of learning? The answer is very simple: accord a higher priority in terms of incentives and regulations to 'spirals of continuous improvement' (quality management) approaches over those advocating the standardization of practices, usually via 'codes of good practices'. For an agronomist, agricultural codes of good practices – the most frequently used tool to encourage farmers to adopt virtuous practices – seem to be contradictory at several levels: (1) they are intended to standardize practices, but often severely restrict the abilities of farmers to adapt to the diversity of soils, climates and agricultural situations; (2) they are codified at the level of the basic agricultural technique, whereas environmental impacts often depend on interactions between several techniques; and (3) they are seen as restrictions, thus devaluing the protection of the environment in the eyes of the farmers. A regulation based on an obligation of results – in contrast to one based on an obligation of means – seems to have an educational value because it urges farmers to conduct diagnoses of their situations (comparison between the obtained 'result', gauged by an indicator provided by the authorities, and the real 'result') and encourages them to implement virtuous cycles of continuous improvement (Meynard, 2010).

The implementation of environmental policies relies increasingly on approaches coordinated between heterogeneous actors, who, even though faced with incomplete information, attempt to adapt the proposed framework to territorial specificities. But these mechanisms are expensive and are applicable only to limited areas. We suggest that, in addition to these mechanisms, governments should encourage enterprises with a territorial scope to play a role in coordinating practices. Thus, in France, agricultural cooperatives, which generally operate on clearly defined territories often several thousand square kilometres in area, are essential nodes of decision making: they sell seeds to farmers and could promote disease-resistant varieties, and associations of varieties or species. Cooperatives provide advice that the farmers respect and, due to their collection function, could encourage crop diversification. Agri-environmental policies, today mainly focused on farmers, should probably also include these enterprises in their ambit, as also local authorities which contribute to the coordination of land use.

5.5 Conclusion

In his report to the UNO on the right to food, De Schutter (2010) emphasizes the urgent need of *'making reference to agroecology and sustainable agriculture in national strategies for the realisation of the right to food.'* To do so, he recommends *'reorienting public spending in agriculture by prioritizing the provision of public goods, such as extension services, rural infrastructures and agricultural research, and by building on the complementary strengths of seeds-and-breeds and agroecological methods.'* The analyses that we have undertaken of input-intensive agriculture and the possible ways of changing it lead to similar conclusions. However, by focusing on the coherence of socio-technical systems and on processes of technological lock-in, we show that the change we desire can only come about if all the stakeholders concerned – first and foremost the government – think and act in a systemic manner. Attractive simplifications such as 'one problem, one solution' or 'one government objective, one instrument' have certainly outlived their usefulness. In such a context, the government has to play the major role of mobilizing all the actors. Various instruments can be used to do so: traditional economic instruments (taxation, quotas, prohibitions, rights trading markets, subsidies, etc.), as well as support provided for research and innovation, for education and training, for collective action and coordination of economic actors. Our approach of studying agronomic processes and their linkages to social and economic systems leads to an analysis that converges with those of the other authors of this book on the concept of innovation systems: it is necessary to act on all components of the social-ecological system to undertake and support changes in agriculture. Changing well-entrenched economic and social dynamics, such as we have described in several regions of the world, will require implementation of proactive policies without delay. Furthermore, it is necessary to signal clear intentions and directions to the many stakeholders so that they can undertake the necessary changes. This will require a consistency in public policies and transparency in regulatory changes. Isn't one of the obstacles in France today to the development of sustainable agriculture the fact that public policies themselves lack sufficient consistency and durability (Meynard, 2010)?

References

Albaladéjo, C., 2012. Les transformations de l'espace rural pampéen face à la mondialisation. Annales de Géographie (in press).

Bergez, J.E., Colbach, N., Crespo, O., Garcia, F., Jeuffroy, M.H., Justes, E., Loyce, C., Munier-Jolain, N. and Sadok, W., 2010. Designing crop management systems by simulation. European Journal of Agronomy, 32, 3-9.

Bland, W.L., 1999. Toward integrated assessment in agriculture. Agricultural Systems, 60, 157-167.

Bockstaller, C., Guichard, L., Makowski, D., Aveline, A., Girardin, P. and Plantureux, S., 2008. Agri-environmental indicators to assess cropping and farming systems. A review. Agronomy for Sustainable Development, 28, 139-149.

Botta, G.F., Tolón-Becerra, A., Lastra-Bravo, X. and Tourn, M.C., 2011. A research of the environmental and social effects of the adoption of biotechnological practices for soybean cultivation in Argentina. American Journal of Plant Sciences, 2, 359-369.

Butault, J.P., Dedryver, C.A., Gary, C., Guichard, L., Jacquet, F., Meynard, J-M., Nicot, P., Pitrat, M., Reau, R., Sauphanor, B., Savini, I. and Volay T., 2010. Ecophyto R&D, quelles voies pour réduire l'usage des pesticides. Synthèse du rapport d'étude. INRA éditeur, France, 90 pp.

Castella, J.C., Jourdain, D., Trébuil, G. and Napompeth, B., 1999. A systems approach to understanding obstacles to effective implementation of IPM in Thailand: key issues for the cotton industry. Agriculture, Ecosystems and Environment, 72, 17-34.

Chevassus-Au-Louis, B., 2009. Refonder la recherche agronomique. In: Les defis de l'agriculture mondiale au XXI ème siècle. Recueil des leçons inaugurales du groupe ESA, Angers, France, pp. 193-226.

Collectif ComMod (Companion Modelling, CIRAD-INRA-IRD), 2006. La modélisation comme outil d'accompagnement. Natures, Sciences et Sociétés, 13 (2), 165-168.

Cowan, R. and Gunby, P., 1996. Sprayed to death: path dependence, lock-in and pest control. Economic Journal, 106(436), 521-43.

Daniel, K., Chatellier, V. and Chevassus-Lozza, E., 2008. Localisation des productions agricoles dans l'UE. L'enjeu de l'évolution des politiques agricole et commerciale. Chambres d'Agriculture 969, 24-27.

Darré, J.P. (ed.), 1994. Pairs et experts dans l'agriculture. Dialogues et production de connaissances pour l'action. Ed. Erès, Ramonville Saint-Agne, France.

David, P.A., 1985. Clio and the economics of QWERTY. American Economic Review, 75 (2), 332-337.

De Schutter O., 2010. Rapport du Rapporteur spécial sur le droit à l'alimentation. United Nations Organization, General Assembly of 20 December 2010, New York, NY, USA.

Dorel, M., Tixier, P., Dural, D. and Zanoletti, S., 2011. Alternatives aux intrants chimiques en culture bananière. Innovations Agronomiques, 16, 1-11.

Dosi, G., 1988. Sources, procedures and microeconomics. Effects on innovation. Journal of Economic Literature, 26 (3), 1120-1171.

Etienne, M. (ed.), 2010. Companion modelling. A participatory approach to sustainable development. Quae, Versailles, France.

Francis, C., Lieblein, G., Gliessman, S., Breland, T.A., Creamer, N., Harwood, R., Salomonsson, L., Helenius, J., Rickerl, D., Salvador, R., Wiedenhoeft, M., Simmons, S., Allen, P., Altieri, M., Flora, C. and Poincelot, R., 2003. Agroecology: the ecology of food systems. Journal of Sustainable Agriculture, 22, 99-118.

Griffon, M., 2009. Pour des agricultures écologiquement intensives. In: Les defis de l'agriculture mondiale au XXI ème siècle. Recueil des leçons inaugurales du groupe ESA, Angers, France, pp. 169-192.

Grosso, S., 2011. Transformations du conseil agricole en région pampéenne argentine et recomposition de la profession 'd'ingénieur agronome'. Thèse Univ Toulouse le Mirail, France.

Labarthe, P., 2010. Services immatériels et verrouillage technologique. Le cas du conseil technique aux agriculteurs. Economies et Sociétés, 44, 173-196.

Lamine, C., Meynard, J.M., Bui, S. and Messéan, A., 2010. Réductions d'intrants: des changements techniques, et après? Effets de verrouillage et voies d'évolution à l'échelle du système agri-alimentaire. Innovations Agronomiques, 8, 121-134.

Lamine, C., Meynard, J.-M., Perrot, N. and Bellon, S., 2009. Analyse des formes de transition vers des agricultures plus écologiques: les cas de l'agriculture biologique et de la protection intégrée. Innovations Agronomiques, 4, 483-493.

Lavelle, P., Blouin, M., Boyer, J., Cadet, P., Laffray, D., Pham-Thi, A.T., Reversat, G., Settle, W. and Zuily, Y., 2004. Plant parasite control and soil fauna diversity. Comptes Rendus Biologies 327, 629-638.

Meynard, J.M., 2008. Produire autrement: réinventer les systèmes de cultures. In: Reau, R. and Doré, T. (eds.) Systèmes de culture innovants et durables. Editions éducagri, Dijon, France, pp. 11-27.

Meynard, J.M., 2010. Réinventer les systèmes agricoles: quelle agronomie pour un développement durable? [Reinventing agricultural systems: what type of agronomy for a sustainable development?]. In: Bourg, D. and Papaux, A. (eds.) Vers une société sobre et désirable. Presses Universitaires de France, Paris, France, pp. 342-363.

Meynard, J.M. and Girardin, P., 1991. Produire autrement. Courrier de la Cellule Environnement de L'INRA, 15, 1-19.

Meynard, J.M., Dedieu, B. and Bos, A.P., 2012 Re-design and co-design of farming systems. An overview of methods and practices. In: Darnhofer, I., Gibon, D. and Dedieu, B. (eds.) Farming Systems Research into the 21st century: the new dynamic. Springer, Berlin, Germany, pp. 407-432.

Meynard, J.M., Dupraz, P. and Dron, D., 2003. Grande culture. Dossiers de l'environnement de l'INRA 2003, (23 Expertise Collective ATEPE: Agriculture, Territoire, Environnement dans les Politiques Européennes), pp. 69-91.

Mischler, P., Lheureux, S., Dumoulin, F., Menu, P., Sene, O., Hopquin, J.P., Cariolle, M., Reau, R., Munier-Jolain, N., Faloya, V., Boizard, H. and Meynard, J.M., 2009. Huit fermes de grande culture engagées en production intégrée réduisent les pesticides sans baisse de marge. Le Courrier de l'Environnement, 57, 73-91.

Pahl-Wostl, C., 2005. Actor based analysis and modelling approaches. The integrated Assessment Journal, 5, 97-118.

Papy, F. and Torre, A., 2002. Quelles organisations territoriales pour concilier production agricole et gestion des ressources naturelles? Etudes et Recherches sur les Systèmes Agraires et le Développement, 33, 151-169.

Rossing,W.A.H., Meynard, J.M. and Van Ittersum, M.K., 1997. Model-based explorations to support development of sustainable farming systems: case studies from France and the Netherlands. European Journal of Agronomy, 7, 271-283.

Sadok, W., Angevin, F., Bergez, J.E., Bockstaller, C., Colomb, B., Guichard, L., Reau, R., Doré, T., 2008. Ex ante assessment of the sustainability of alternative cropping systems: implications for using multi-criteria decision-aid methods. A review. Agronomy for Sustainable Development, 28, 163-174.

Schott, C., Mignolet, C. and Meynard, J.M. 2011. Les oléoprotéagineux dans les systèmes de culture: évolution des assolements et des successions culturales depuis les années 1970 dans le bassin de la Seine [Oilseeds in cropping systems: changes in rotations since 1970 in the Seine water basin]. Oléagineux, Corps gras, Lipides, 17, 276-291.

Soulard, C.T., 2005. La multifonctionnalité de l'agriculture en pratique: étude des relations entre exploitations agricoles et étangs de la Dombes. Cybergeo, European Journal of Geography, 319. DOI: 10.4000/cybergeo.6610.

Thenail, C., Joannon, A., Capitaine, M., Souchere, V., Mignolet, C., Schermann, N., Di Pietro, F., Pons, Y., Gaucherel, C., Viaud, V. and Baudry, J., 2009. The contribution of crop-rotation organization in farms to crop-mosaic patterning at local landscape scales. Agriculture Ecosystems and Environment, 131, 207-219.

Vanloqueren, G. and Baret, P., 2008. Why are ecological, low-input, multi-resistant wheat cultivars slow to develop commercially? A Belgian agricultural 'lock-in' case study. Ecological Economics, 66, 436-446.

Vanloqueren, G. and Baret, P., 2009. How agricultural research systems shape a technological regime that develops genetic engineering but locks out agroecological innovations. Research Policy, 38, 971-983.

Warner, K.D., 2007. Agroecology in action; extending alternative agriculture through social networks. The MIT Press, London, UK.

Wezel, A., Bellon, S., Doré, T., Francis, C., Vallod, D. and David, C., 2009. Agroecology as a science, a movement or a practice. A review. Agronomy for Sustainable Development, 29, 503-515.

Chapter 6. Innovation and social inclusion: how to reduce the vulnerability of rurals?

Denis Requier-Desjardins

6.1 Introduction

Innovation is currently at the centre of the discourse on the paths of development and growth in both the North and the South. This manifests most often in the form of repeated references to the development of the 'knowledge economy', based on a range of information and communication technologies, and its disruption of existing 'technological paradigms' in at least as significant a manner as the impact of the first industrial revolutions on the world economy.

The question of innovation concerns therefore all the countries of the South, regardless of their level of development: their integration into the knowledge economy has become a fundamental issue. It also concerns agriculture and the agrifood sector for at least three reasons:

- First of all, agriculture and the agrifood sector depend almost exclusively on using living organisms in their production processes. And certainly biotechnology is one of the most dynamically innovative sectors of our time, with a number of major breakthroughs whose impacts are, as yet, not fully assessed.
- Secondly, precisely because of their use of living organisms, agriculture and the agrifood sector are areas with severe constraints where the urgent requirement of creating and adopting sustainable development processes finds full expression and drives current innovation efforts. Issues relating to biodiversity, climate change and land use – as debated in the three major conventions on the environment – highlight this very clearly. Obviously, agriculture is at the forefront of this effort.
- And, finally, the ability of agriculture to feed a growing world population (even if the rate of growth is slowing) is highly conditional on sustainability, particularly for vulnerable populations whose access to food is not guaranteed. These populations are often very closely connected to agriculture since global poverty remains overwhelmingly rural. The issues of development at the beginning of this new millennium are increasingly identified with the fight against all types of poverty, monetary as well as that of capabilities, and

against poverty-related discrimination. They also encompass a search for paths of sustainable development, themselves having a poverty eradication component. Witness, for example, the definition of the 'Millennium Development Goals'.

It is therefore legitimate to consider the issue of innovation in agriculture and the agrifood sector in terms of its ability to reduce the incidence of poverty in rural areas of the South and to promote the emergence of a path of sustainable development which includes not only the environmental dimension of sustainability but also its economic and social aspects. It is this 'hot' issue that this chapter is devoted to.

In this perspective, we advance the hypothesis that innovation's role in the fight against poverty in the South should be judged differently when we consider the reduction in poverty of family farmers alone as opposed to quite simply the reduction of overall rural poverty. If, in the first case, we can define it in the form of agricultural innovations which can be adopted by small farmers and which can boost their incomes by increased production efficiency, in the second, it has to refer to all rural activities and practices and to decision-making processes within the household; it therefore no longer remains specifically agricultural in scope. The definition of sustainable rural development paths in the South is thus modified. The issue that then arises is that of the relationship between these two approaches to innovation in rural areas.

- We will take a three-stage approach to addressing this issue. We will start by reviewing the concepts of innovation, poverty, rurality and the many related issues that arise, particularly those concerning the economics of innovation and development. We will cover the ongoing debate on the relationship between agriculture and rural life.
- Secondly, we will focus on innovation processes that affect agriculture and the agrifood sector.
- Finally, we will examine innovative practices implemented by rural households and, on a wider scale, innovation in rural territories.

In conclusion, we will discuss national and international public policies for fighting poverty within this context.

6.2 A review of certain concepts

First we turn briefly to the concept of innovation before addressing the notions of poverty and its relationship to inequality, and of rurality and its relationship to agriculture.

6.2.1 Innovation

Initial thinking on innovation (Schumpeter, 1911) revealed a typology of innovations which goes beyond technological innovation focused on just the production processes. Also included were product innovation, organizational innovation and market innovation. Indeed, any serious discussion on innovation cannot be limited to the strict framework of production processes but should take into account organizations, institutions and social practices – even, when necessary, from an economics standpoint.

In a 'neo-Schumpeterian' perspective, the economics of technical change and, in particular, 'evolutionary' approaches (Nelson and Winter, 1982) have highlighted the distinction between disruptive innovation and incremental innovation, a distinction that refers to the notions of technological paradigms and technological trajectories: A disruptive innovation creates a new technological paradigm, i.e. a set of principles that guide the search for new processes, whereas incremental innovations develop this paradigm without actually questioning its underlying principles. The innovations are indeed dependent on the path, each limiting future options and locking-in the technological trajectory, which can only be challenged by a subsequent disruptive innovation. Though initially developed around technological innovations, these concepts can be extended to other forms of innovation, especially to organizational and institutional ones.[1]

This stream of analysis also proceeded to broaden the scope of identifying actors of innovation. Schumpeter made a distinction between invention and innovation: the invention is a process outside the marketplace, taking place upstream of innovation, whereas innovation consists of the economic valorisation of an invention by an entrepreneur, innovation's key actor. The later phase of disseminating the innovation was not intended to affect the content of the innovation. Yet evolutionary economics reintegrates the invention in the innovation process and considers research activity as an economic one. Thus the entrepreneur or the company is no longer the only actor of innovation. Innovation becomes a complex process with feedback loops (Kline and Rosenberg, 1986) involving users of innovation, companies or civil-society actors, who may intervene in the process, notably by contributing incremental improvements. Innovation can therefore be perceived as a product of collective action, particularly through the identification of innovation networks, linked to forms of knowledge. 'Knowledge economy' refers to these complex processes: knowledge is both a non-rival good and an appropriable good, notably via mechanisms of intellectual property. Furthermore, knowledge is its own input. The innovation system is then faced with a tension between, on the one hand, the need to create and guarantee ownership rights over the innovation to ensure compensation of the cost of innovation and, on the other, the necessity of disseminating it as widely as possible to benefit from the feedback loops mentioned above and to foster the emergence of new innovations.[2] Agriculture is obviously not immune to these processes; it relies on the use of living organisms and therefore on biotechnology, the subject of frequent intellectual-property

[1] North (1990) has most notably shown that, at the scale of nations, institutions were dependent on the path of institutional innovation, which can engender the phenomenon of institutional lock-in. This calls into question the position he had originally advocated of an optimization of the institutional trajectory in the context of the development of market economies.

[2] Debates on intellectual property echo this tension, for example, through the issue of patent term limits or the possibility of recognizing a collective intellectual property.

disputes concerning the 'patenting of life'. More generally, innovation processes in agriculture will bring into play feedback loops involving actors far beyond agriculture *per se.*

6.2.2 Approaches to poverty and inequality

Poverty is traditionally approached with reference to monetary income and a 'poverty line' representing an income level that can satisfy a consumption level considered essential for a given society. However, this approach which, in any case, encounters several problems in defining the poverty line (national or international, absolute or relative as a percentage of the median income, for example), has been challenged by the development of a 'capabilities' approach, first proposed by Sen (1987, 1999).

If we restrict ourselves for the time being to the income and poverty line approach, the simple measure of the incidence of poverty (the proportion of the population below the poverty line, the most common indicator) has to be complemented by considering the intensity of the poverty (the mean deviation from the poverty line), as well as the depth of poverty (the distribution of poverty amongst the poor). The inclusion of the latter two metrics allows the identification of extreme poverty and to assess the effects of selection among the poor. This aspect becomes even more significant when the poor are not only the objects of specific poverty-alleviation policies but are additionally affected by so-called 'pro-poor' sectoral or macroeconomic policies where poverty reduction is but one of several goals of a specific policy. Such policies may, in fact, exacerbate selection effects amongst the poor. Their specific impacts depend mainly on the intensity and depth of poverty of the population concerned.

However, the use of a strictly monetary-related measurement of poverty, on which the definitions of the indicators we have just mentioned are based, as an exclusive approach to this phenomenon has been widely challenged:

- On the one hand from the beginning of the 1980s, by the 'basic needs' approach which identifies poverty by the lack of direct access to a number of goods and services, in general with a strong public nature and partly off-market (water, health, hygiene, education, etc.). The level of access to these 'basic needs' is then assessed by social indicators of development which do not refer to a monetary measurement.
- And, on the other, by the approach in terms of 'capabilities', which extends and amplifies the basic-needs approach. It substitutes 'good life' for simple well-being as a criterion of personal situations. Well-being refers to material satisfaction at any given time, 'good life' refers to life taken as a whole in all its dynamics that ensure an individual's self-esteem and social recognition and which also conforms to his or her values. This approach emphasizes the availability of choice in the basic aspects of a person's life, especially as far as health, education, culture and security are concerned (for example Box 6.1). Also included in the concept of well-being is the availability of a number of fundamental freedoms, such as of association and expression, which ensure the recognition of an individual within society.

Box 6.1. Horticulture and the capability approach in Cameroon.
Laurent Parrot, Philippe Pedelahore, Hubert De Bon and Rémi Kahane

The capability approach allows the characterization of transformations taking place in rural sub-Saharan Africa by describing, for example, the links between the emergence of a rural non-agricultural sector and migration patterns. It is therefore a matter of determining the factors that would make it possible for people to achieve various lifestyles.

According to Dubois and Mahieu (2002), the building of these capabilities depends on three factors: (1) goods and potential possessions which include available capital, various assets, social relationships, beliefs, etc., (2) personal characteristics and (3) social opportunities, in particular the position of women in society.

These three factors were studied in Cameroon for an improved understanding of the emergence of horticulture as primary agricultural activity of households located in a periurban area at Muea, 40 km from Douala.

We have shown that horticultural activity requires various assets such as access to land and to credit for purchasing inputs. It requires, in addition, access to informal social networks to acquire technical knowledge and market information. These factors probably explain why the horticulturists are, on average, older than their staple crop-farming counterparts. We have also established a link between the practice of horticulture and diversification into non-agricultural rural activities, with households relying on their non-agricultural activities to fund horticultural systems. Finally, we have highlighted the preponderance of women's groups specializing in horticulture thanks to their social networks which are especially effective in accessing credit and information.

These various assets are found to be necessary for a successful transformation of agriculture from essentially food-based farming based on few inputs and low productivity and yields (subsistence farming) to a form of agriculture suitable for urban and periurban areas, where intensification processes require specific knowledge (commercial agriculture). Thus, the urban transition could be beneficial by enabling young people to migrate initially to town, accumulate sufficient capital and acquire a certain entrepreneurship, then return with the intention of implementing new production systems or agricultural commercialization.

This approach, integrated in time and space, shows the importance of support policies which seek an improved valorisation of the positive interactions between the urban and rural sectors as part of an integrated-development framework (Parrot *et al.*, 2010).

In addition, the relationship between poverty and inequality has been the subject of several discussions. These have highlighted two different, albeit interlinked, variables that are involved: inequalities reinforce the subjective perception of poverty but the ripple effects of increased spending by those better off can sometimes reduce poverty while, in contrast, certain measures to reduce inequalities can sometimes adversely impact economic growth and thus the population's general standard of living. The Rawlsian theory of the 'veil of ignorance' (Rawls, 1971) aims to resolve this contradiction theoretically by emphasizing the 'lexical' priority of fundamental freedoms and by making the improvement of the lot of the most disadvantaged a criterion which can accept some forms of inequality.

Finally, in the context of poverty-alleviation policies, the importance of targeting, particularly to the poorest, was highlighted. This concern, which refers to the differentiation of poverty situations highlighted above, can sometimes conflict with a general principle which advocates the implementation of so-called 'pro-poor' economic policies. In fact, these policies do not have the central objective of reducing poverty; it is only one of their many goals. The consequence of these policies may be the exacerbation of selection effects among the poor, although, of course, there may be a reduction observed in the incidence of poverty but not necessarily of its intensity or its depth.[3]

Such considerations thus call into question the fairness of pro-poor policies, even though their stated goal is precisely that of bringing about equality within the concerned societies.

6.2.3 Agriculture, the agrifood sector and rurality

Before we can examine the role of innovation in the agricultural sector, we have to be able to define this sector unambiguously. Two traditional characterizations are:

- Agriculture uses processes of living organisms for producing primarily food products and some industrial raw materials.
- Agriculture is practised mainly in rural areas, of low density because of their specific relationship to space and land use.

These two definitions allow us to include agriculture in the wider structures which surround it.

First, agriculture is now highly integrated with the agrifood sector constituting a 'global value chain'[4], mainly dominated by downstream players, the supermarket chains. The question of innovation should thus be asked at the scale of the entire agrifood sector rather than only at the level of strictly agricultural activities, given that network effects and feedback

[3] We note that assessing the impact of innovation policies on poverty alleviation comes down to assessing a mechanism of this type.

[4] As defined by Gereffi (1999) and Gereffi and Humphrey (2005).

loops mentioned above will connect the various links of the value chain, from agricultural production to distribution to the consumer. Not only are the various processing stages included but also agricultural supplies and inputs. Recent work has emphasized, for example, the impact of innovations in the entire chain, especially those relating to standardization, on agricultural producers of the countries of the South and, in particular, on small producers (Reardon *et al.*, 2009).

And second, agriculture refers to rurality, as seen from an economic viewpoint of systems of productive activities practised in rural areas. Both terms have even been considered synonymous since agricultural production is an essential feature of rural life. But, even if the rurality of an area has been long connoted by the extent of agriculture as a dominant economic activity, which even allowed rural households to be equated to 'family farming', we see today a double shift between agriculture and rurality.

- On the one hand, we are witnessing the growth in non-farm activities and incomes – non-agricultural for the most part – in all rural areas, especially in the South (Haggblade *et al.*, 2010; De Grammont, 2010 for Mexico). This diversification first manifests itself at the household level, which is then emboldened to develop a portfolio strategy consisting of this set of activities in order to minimize the risks in the household's 'economic livelihood' (Ellis and Freeman, 2005). Some commentators believe that agriculture can only be a marginal component of this portfolio, a variable of adjustment even (Rigg, 2005).
- The outcome at the mesoeconomic level of rural areas of these situations in terms of households may be a consolidation of the 'residential' economic base, relying on the multiplier effects of externally generated income spent locally (tourism, migration, various types of remittances). This base can sometimes replace a 'productive' one and results from the export of goods and services outside the territory, the latter being primarily agricultural in rural areas. The ripple effects can then stimulate the development of 'local' activities (Davezies, 2008), which are generally non-agricultural in character.
- On the other hand, and perhaps not much attention has been paid as yet to this aspect, forms of agriculture are emerging which can best be described as enterprise or business farming, in which agricultural activity adopts the same organizational model as in many industrial sectors. This is characterized partly by the role of financial capital and the involvement of financial actors in the activity and partly by the mastery of technological mechanisms based on developments in the knowledge economy for, notably, allowing widespread use of outsourcing and the creation of 'network agriculture' (Hernandez, 2008; Clasadonte, 2010). In this type of model, land, which traditionally defined the farm and identity of farmers, becomes just one generic input amongst many others. The model throws up opportunities of global arbitrage, which further disconnects agriculture activity from its territorial moorings and thus from rurality.

These elements lead to a 'new rurality' in which the income of rural residents no longer depends exclusively – or even primarily – on farm agriculture and in which agriculture breaks its connection with the territory. The question to then ask is whether, given this disconnect, innovation in agriculture has an impact on the poverty of rural households as defined in the broadest possible manner. Should we not address directly the issue of innovation in rural areas, instead of the narrower scope of innovation in agriculture? This issue also refers to the type of innovations, given that innovation in agriculture will have a significant technological content whereas rural areas may benefit most from the emergence of innovations that are more social or organizational in nature. An examination of the impact of innovation on reducing rural poverty cannot be limited to innovation that helps the agricultural production sector alone; the entirety of innovation processes that impact rural economies and societies have to be considered.

The rest of this chapter will be devoted to this pertinent issue. We shall first evaluate to what extent agricultural innovation can be a factor in reducing poverty. We shall then address the question of innovation capacities in rural areas and their impact on poverty.

6.3 Is agricultural innovation a factor in the reduction of poverty and vulnerability and in the sustainability of development trajectories?

Agricultural innovation has been particularly dynamic in recent decades, as evidenced by the significant increase in yields and, in some areas, in production per agricultural worker, even if the rate of this increase is now slowing.[5]

Broadly speaking, we can distinguish today between two paths of innovation, relatively divergent – even if they do meet in some contexts – because of the lock-in effect that they are subject to. This divergence gives rise to two different visions of the future of agriculture.

The first path, which can be described as production-intensifying, is an extension of agricultural innovations first proposed in the 1970s to the farmers of the South as the green revolution, i.e. the use of selected seeds, mechanization and agro-chemical inputs (fertilizers and treatments), leading farmers towards specialization and thus towards monocultures. This path is today strongly influenced by advances in biotechnology, particularly as regards seeds, with the emergence and development of genetically modified organisms (GMO). Use of these GMOs leads to reduced costs of field crops by increasing economies of scale in phytosanitary treatments (seeds tolerant to Round-up) or in tillage (wider use of direct seeding). Such a path has a significant impact on the organization of production, usually in the form of changes that promote economies of scale.[6] The lack of these very economies of scale in agriculture was commonly explained by the predominance of family farming structures, especially due

[5] According to the Agrimonde forecast (Paillard et al., 2010) production, as measured in Kcal/day per ha, increased by a factor of 2.5 between 1961 and 2003, with growth being particularly strong in Asia. In the OECD areas, food production per farm worker increased by a factor of 7 but growth was much slower in other regions.

[6] The structuring of agriculture into networks and 'pools de siembra' (network companies) in the Southern Cone of Latin America is the most obvious illustration of this increase in production scale (Hernandez, 2008).

to these farms' high transaction costs of using hired labour (Lipton, 2006) and to their low rate of utilizing physical capital. 'Network agriculture', mentioned above, can today reduce or outsource labour-intensive agricultural work and thus substantially increase acreages under cultivation, and can be seen as a sort of culmination of this path.

In the framework of this intensification path, environmental sustainability can only rely on a 'segregationist' approach which proposes that ecosystems can be managed independently of agriculture, notably by the establishment of protected areas devoid of any agricultural activity.

The innovations offered to small and family farmers in the South have long been part of this paradigm: promoting access by small farmers to inputs and mechanization, mainly via subsidies. A bias, however, exists towards innovations for food crops, to the detriment – in form of lack of research funding, for example – of traditional cereals and tubers. It is also believed that innovation should be adapted to small farms, corresponding roughly to family farming, since it constitutes the bulk of the agricultural labour force. Moreover, yields are generally low in this sector and thus have a potential for showing considerable improvement. In this way, innovation can have an immediate impact on the income of farmers, usually the first victims of poverty. Emphasis is laid in this vision, for example, on agricultural extension policies in combination with policies of access to 'missing markets' (credit, inputs, etc.) and to commercial markets for selling produce (transport infrastructure). Therefore technological innovations must be complemented by organizational innovations and by infrastructure investment for improved access to markets and thus for an increased scale of production. For some export-destined products, mainly in the area of non-traditional exports (vegetable and fruits, flowers), forms of contract farming – based on controlled use of irrigation and inputs as well as strict compliance with quality standards defined by dominant actors in the supply chains – can also apply to family farmers.

But the overall result of this intensification proposed to the small farmers can seem disappointing (Dethier and Effenberger, 2011). At best it generates a selection effect: only a minority of small farmers can integrate themselves into these systems and eventually upgrade their functions and skills.

The second path, which appears as a disruptive paradigm, originates from questions on the environmental impact of agricultural practices, especially ones following the first path which intensively use inputs and become highly specialized. This second path forms part of the debate on the multifunctional character of agriculture and on its ability to 'manage cultivated ecosystems' while preserving them. It proposes an 'agro-ecological' intensification which encourages polyculture rather than monoculture, natural fertilizers over chemical ones and biological control instead of recourse to treatment products. It features innovations in agricultural activity, particularly in terms of crop rotation and intercropping, which preserve the ability of the ecosystem to regenerate itself sustainably while helping to increase food production at minimal environmental cost. These types of innovations can also be combined with innovations elsewhere in the food-supply chain which minimize the environmental impact, for example, by reducing transport costs by attempting to develop commercial 'short supply chains' based on agreements between producers and consumers. This path recalls the 'doubly green revolution' (Conway, 1997; Griffon, 2006) which would replace the 'green revolution' approach of the 1970s for the development of agriculture in the South.

While these innovations can, of course, be implemented on large farms, the focus of this path of innovation remains the various forms of family farming. These farmers are, *a priori,* more likely to be affected by poverty or just by the vulnerability of their labour systems. Indeed, this path includes elements (intercropping and crop rotation, use of adapted traditional varieties) that are considered to form part of traditional agriculture's cognitive heritage. We are therefore in the presence of a new formulation of the theory of comparative advantage of family farming, based less on the transaction costs associated with the use of hired labour and more on its adaptation to the constraints of multifunctional agriculture and on its respect for the environment.

As far as the impact on poverty and vulnerability of the rural or urban poor in countries of the South is concerned, the first expected impact of innovations in agricultural intensification is an increased agricultural productivity and hence improved yields. This increase in the productive capacity of small farmers is a factor in the reduction of their exposure to poverty in the context of the 'three ways out of poverty' vision of the World Bank (2008) and it should lead to a logical increase in food availability. However, the question that should be asked is whether such an increase in itself guarantees improved food security of the concerned populations, particularly in countries of the South (Delthier and Effenberger, 2011).

In fact, this type of approach, which focuses primarily on the relationship between production and food supply, has been called into question as far back as in the 1980s in the debate on the foundations of 'food security'. Sen's (1982) critique of the hypothesis of FAD (food availability decline) in explaining the causes of famines and food crises played a fundamental role in this paradigm shift. This change culminated in the highlighting of the importance of an individual's 'right to food'. Increased production is not in itself a guarantee of household food security; it is the household's ability to obtain food by the exchange of rights which is crucial, and food insecurity is but one aspect of general household vulnerability.

Nevertheless, the reference to food sovereignty can make agricultural productivity in a territory – especially when considered at the national scale – a major factor in food security since it can help limit price shocks, notably by supplying urban areas. Geographical or even organizational proximity between producers and consumers, for example in the case of 'short supply chains', is also a factor in reducing risks, both for consumers and producers. In such a case, the local food security of households depends primarily on local production capacities and the degree of organization of agricultural producers. Organizational innovations in the food-supply chains, particularly those that strengthen the links between producers and consumers, can then be considered as factors that help fulfil the 'rights to food'. Such organizational innovations are consistent with a path of innovation which focuses on environmental sustainability based on agro-ecological innovations.

In addition, one must consider not only agricultural innovations but also those in the agrifood value chains. A large body of literature has focused on the transformation in governance of these sectors in the context of globalization and the subsequent impact on family farming. There has been a rise in the dominance, on the one hand, of supermarkets in developing countries and, on the other, of specialist wholesalers supplying specific products to the centralized purchasing departments of the supermarkets (Reardon *et al.*, 2009; Reardon and Timmer, 2007; etc.). The consequence is that domestic markets in which family farmers are present are often, in fact, already globalized. They operate on the basis of technical and organizational innovations such as the use of information technology or conformity with quality and processing standards that the farmers are required to meet. Some family farmers can integrate themselves into these value chains but the selection effect between farmers can be brutal. For example, a substantial gap can develop between a group of farmers who are under contract and others using traditional market channels in decline (wholesalers, transporters and wholesale markets, for example). This latter group may see their agricultural production activities become marginalized.

This path of innovation in agriculture and the agrifood sector highlights some characteristics of innovation specific to other sectors, most notably the importance of innovation networks and collective action, and the importance of processes of exclusion that may accompany innovation.

Even though the literature advances the notion that the conditions for innovation in agriculture may be different, especially as this sector is less amenable to economies of scale than others and because it is based on the mechanisms of living organisms and on knowledge accumulated over generations of farmers, it seems that agriculture is increasingly seeing innovation processes which are, in fact, quite similar to those found in other sectors. For example, many studies have explored the role of networks in the innovation process, especially within clusters. It is significant that the seminal studies on the topic, not limited to any particular sector, have chosen to use agricultural examples (Guiliani, 2007, on the Colchagua wine cluster, for example, in the case of network analysis or Porter, 1990, on the California wine cluster.)

The issue of innovation networks relates to that of collective action in agricultural innovation. It also refers to the question of ownership of innovations and, in particular, to that of intellectual property. Two aspects are of prime importance in agriculture: (1) the existence of patents that relate to the 'patenting of life' and, incidentally, the duration of such patents, and (2) the recognition of traditional knowledge of the populations concerned as collective intellectual property. We are, however, confronted with a general question: how to decide between the necessary recovering of the costs involved in the innovation processes and the no less necessary widespread dissemination of innovations? Indeed, the processes of exclusion resulting from networks of collective action enhance the effectiveness of this action but also exacerbate inequality and exclusion, as shown sometimes in countries of the South by the example of quality labels.

In conclusion, innovations for intensifying agriculture risk creating effects of selection and contributing to the marginalization of some sectors of the rural population, especially small farmers. Obviously, we cannot ignore the impact that these selection processes may have on reducing the incidence of rural poverty by improving the standard of life for many family farmers. Nevertheless, the issue of their impact on the intensity and depth of poverty still remains, given the effects of marginalization of those who cannot join the networks of innovation. It should be acknowledged that there are other ways out of rural poverty, in particular by the diversification of rural household activities. The question of the formation of networks and their territoriality as well as those concerning the organization of producers or local agreements between producers and consumers can be reframed in terms of the relationship between agriculture and rurality via their reference to the territory.

Moreover, the agro-ecological innovations destined for agricultural producers are often proposed to them under a 'project cooperation' framework by organizations and defined by experts or technicians. These innovations do not necessarily conform to the value systems or practices of the populations concerned (Dietsch and Ruault, 2010) and it is often difficult to judge whether they can be easily adopted by small farmers. Most often, the proposed innovations focus on agriculture and are based on the assumption that the only way out is to boost agricultural production. The selection of leaders for promoting these innovations already introduces to some extent the specific effects of selection associated with the innovation mechanisms of the dominant path.

We should perhaps then explore innovations that can directly reduce rural poverty, innovations which are not necessarily agricultural in character. Indeed, the development of income-generating non-agricultural activities in rural areas is equally likely to diminish the impacts of poverty. These activities can also be linked to migration, which can appear to be one 'way out of poverty' (World Bank, 2008). From a wider perspective, innovation can be seen as a process with which a human group uses knowledge produced by various agents to create new value (Fernandez-Baca *et al.*, 2010). When applied to a rural community, this definition broadens the scope of innovation beyond that of simple agricultural activity.

A participatory approach can perhaps help select innovations for reducing rural poverty where adequate participation by the concerned populations would accord legitimacy to these innovations. However, in such a case, it is necessary to prevent the misuse of the 'participation' as a method of project management since selection effects can be enhanced and the identification of target groups may become based only on *a priori* representations of rural society.

6.4 Innovations for reducing rural poverty

What concerns us are innovations, technical as well as organizational or institutional, that can help combat poverty and vulnerability of rural households, keeping in mind that rural households do not necessarily equate to units of family agriculture and could even have nothing whatsoever to do with agriculture.

If we consider the question of reducing poverty in rural areas in this manner, i.e. by excluding technical or organizational innovations that only concern agricultural production processes, we are left only with organizational and institutional innovations that can help improve access to food and to other components of well-being. The issue resolves around the proponents of innovation, who need no longer be limited to farmers or public or private institutions involved in the processes of technical innovation in agriculture.

Indeed, we can start by assuming that these innovations, which intend, for example, to define new strategies for increasing income are initially implemented by the very economic units that will benefit from them, i.e. the households themselves, via their practices and their choice of activities. But we can envisage also collective-action approaches at the local, territorial level. Finally, we can see innovations originating with and driven by appropriate public policies. We explore all these three levels below.

6.4.1 Can rural households develop innovations to reduce their exposure to poverty and vulnerability?

The poverty and vulnerability of rural households are reflected by their standards of living considered in combination with their ownership of different forms of 'capital' (Ellis and Freeman, 2005). We can normally distinguish between physical or productive capital, land capital, human capital, financial capital and social capital. These tangible and intangible assets may be invested in a range of activities not only to maintain the household's level of well-being but, more importantly, also to withstand shocks that may affect its livelihood. The concept of risk management is central to these strategies. A household increases its resilience to shocks by using a strategy to manage its portfolio of activities. It is this that will determine whether it can emerge from poverty and vulnerability or not. Innovation seen at this level thus refers to emergence of new activities and new sources of income from the same existing portfolio of household assets.

Can we therefore consider strategies for distancing rural households from agriculture activity, such as diversification of activities or even migration, as innovative strategies for poverty alleviation? We can consider the example of migration, a phenomenon that has become significant in some rural areas (Requier-Desjardins, 2010).

At first glance, it may seem surprising to consider the various forms of migration prevailing in rural areas, temporary or permanent, national or international, as innovative strategies. Nevertheless, existing literature on the subject of migrations and the importance of migrant remittances in the economies of rural areas has highlighted the following two points:

- First, faced with initial studies which highlighted the 'push' aspect of migration, perceived as the result of totally constrained decisions, studies on the new economics of labour migration as also sociological work on migration networks show that migration is primarily a family decision whose aim is to optimize the use of household resources. More broadly, the UNDP 2009 Human Development report devoted to migration highlights, from a perspective of 'capabilities', the fact that mobility is a human right and that migrants are primarily expressing a project of life by migrating.
- Second, the debate on the use of migrant remittances in rural areas of the South, while very controversial, not only highlights a relative consensus on their contribution in poverty alleviation but also points to a strengthening of human capital of populations which benefit from it. This in itself is an innovation-friendly factor. The issue of financing of productive investments and thus of development of new activities, even of technology transfers, is still being debated but some aspects seem to show that remittances can be a source of infrastructure building, of financial development, and can even facilitate the adoption of new technologies, for example, in the irrigation domain. The circulatory character of migration can play an important role in this regard.

The highlighting of these points in the literature has led to somewhat radical positions: Rigg (2005) argues that the fight against poverty in rural areas should, at least in many territories in the South, be disconnected from the issue of agricultural development. By aggregating migrant incomes with those of residents, Clements and Pritchett (2008) consider migration not as an alternative to development but as one of its constituents in the countries concerned.

Such reasoning does not limit itself to migration's various forms but also considers the diversity of activities outside of agriculture. These activities call for some household members to acquire new skills. For example, numerous experiments have been documented on the development by women's collectives of food and non-food artisanal activities (Boucher *et al.*, 2010; Box 6.2).

6.4.2 Territorial innovations at the mesoeconomic level

On a more mesoeconomic level, organizational innovations play a key role, especially in terms of governance of rural areas. They rely in particular on processes of collective action,

Box 6.2. LAS (Local AgriFood System): a new tool for the development of marginal territories. Lessons from the Rural Agro-Industries Alliance of Selva Lacandona, Chiapas.
Francois Boucher, Denis Requier-Desjardins and Virginie Brun

The geographical concentrations of RAIs (Rural Agro-Industries), most often observed in Latin America, give a strong territorial identity to these local-development dynamics, which rely on complex interactions between territory, actors, products and innovation systems (Boucher *et al.*, 2003).

Based on this observation, a project to develop rural micro-enterprises of type RAI was launched in several Indian communities of Selva Lacandona at Chiapas, Mexico. This is a region that is among the most isolated, marginalized and poor in the country. The project's objectives were threefold: reduce poverty in several microregions of Selva Lacandona; reduce pressure on natural resources and slow down environmental degradation processes; and integrate social development policies in a framework of participatory and sustainable territorial development. Two broad lines of action were defined on the basis of these objectives: on the one hand, helping organize and launch sixteen RAIs of various types (food processing, handicrafts, water purification micro-plant, sales cooperative, etc.) and, on the other, strengthening management and innovation capacities of those involved. Different participatory workshops were thus organized around themes identified as priorities: technological innovation, organization and management of an enterprise, improvement in production processes, market analysis and marketing. This first stage was then used to induce a collective territorial dynamic around the consolidation of these various RAIs and the dissemination of knowhow and innovation. These activities were formalized by the setting up of an organization, 'Rural Agro-Industries Alliance of Selva Lacandona', and the creation of a collective brand, 'Rural Agro-Industries of Selva Lacandona'.

One of the key lessons learned from this development project was the ability to identify and define conditions for the viability of RAIs in Selva Lacandona. Even though economic profitability of micro-enterprises may seem essential for their survival, it is not a central issue since it does not pose any real problems. But the two factors that did emerge as fundamental to the sustainable development of RAIs in a marginalized region such as Selva Lacandona were the necessity of a favourable and enabling environment and the resolution of organizational and leadership issues. The first point, already addressed by De Janvry and Sadoulet (2002), in their three-stage approach to poverty reduction, refers to the necessity of investments in primary services (education, health, nutrition, infrastructure and other basic services). They are indeed prerequisites for the creation and sustainability of small rural enterprises. It became apparent that a dynamic like a LAS, based on the horizontal dissemination of innovation processes and on territory-specific know-how and resources, is intrinsically linked to the presence and use of local and functional public benefits – which are usually lacking in marginalized regions. The second point refers to the problems of organization, collective action and leadership within groups, most often the result of distrust between members and of roles not clearly understood. The main result is the blocking of the collective dynamics, as manifested by the presence of 'free riders' within groups and by situations where the leaders adopt a 'dog-in-the-manger' attitude, neither doing nor allowing to be done (Boucher *et al.*, 2010).

such as the strengthening of local educational or health infrastructures or the development of marketing strategies to improve producer incomes. In this category, we can mention, in particular:

- Strategies relating to the establishment of fair-trade mechanisms. An illustrative example is the implementation of approaches for marketing fair-trade coffee, which involves the creation of farmer collectives and the investment of part of the income in building up community infrastructures.
- Innovative marketing strategies, including the development of short supply chains and the establishment of proximity ties with consumers. They tend towards a territorialisation of agricultural production and a sharing of production risks with consumers. Even though this type of strategy was developed mainly in the North (witness AMAP in France), a few experiments have taken place in the South, especially in Latin America, for example the creation of farmers' markets, organic markets or short supply chains serving cities (Sao Paulo).

In some contexts, such as that of the Andes studied by Fernandez-Baca *et al.* (2010), we find that community innovation projects reflect the diversification of activities and the corresponding portfolio of assets (Box 6.3). The importance of artisanal and tourism projects in rural communities in countries of the South should also be emphasized.

However, by going beyond these approaches, we can expand the scope of innovation at the territorial level by concentrating on the territory's 'specific assets' and by establishing a strategy of territorial qualification.

From this perspective, innovation strategies for qualifying territories do not revolve around just agricultural products. A 'basket of goods' (Pecqueur, 2001) approach can reveal links between different activities and various goods and services, private or public, which share a territorial-qualification characteristic[7]. This characteristic is defined notably by an interaction with consumers who validate it and it can be strengthened by institutional mechanisms that result from collective action. At the territorial level, this characteristic can, especially in the context of diversifying rural activities, promote a synergy between agriculture and other activities, for example via the creation of 'local agrifood systems' built around high quality products. The sustainability of such a path of territorial development, however, requires some stabilization in demand for the basket of goods. But this demand is largely driven by various forms of residential and 'presential' bases, built around available territorial amenities. As an example, one has just to consider the role of tourism or that of the migrant diaspora, which can be equally important. We can thus raise the question of replicating this type of trajectory. What will be the applicability in rural areas of which countries? Moreover, considering innovations at the territorial level – and not just at that of the household – does permit us to raise the issue of the scope and nature of policies and projects which can be defined at this level.

[7] As defined by Lancaster (1966).

Box 6.3. Contributions to the promotion and development of rural innovations: lessons from 'Panorama Andino' on rural innovation in the Andes.
Edith Fernandez-Baca , Maria Montoya, Natalia Yanez

The Consortium for the Sustainable Development of the Andes Ecoregion (CONDESAN) conducted a study on the state of the art on rural innovation in the Andes region. Twenty cases were documented in rural Andean areas in seven countries (Argentina, Bolivia, Chile, Colombia, Ecuador, Peru and Venezuela).

This study found that though all countries face development and economic growth problems, these vary widely and thus each country uses different types of instruments to address their economic, social and environmental realities. Development and growth policies, as well as policies for poverty reduction, are also different and heterogeneous among countries.

From an endogenous perspective, this heterogeneity can be considered a strength, since it is based on local and/or national capacities and competencies, and on the degree of institutional development that each segment of the population has. However, it can also impede the establishment of common innovation agendas for the Andes region. Nevertheless, it is possible to learn how the capacity to innovate develops in diverse scenarios of scarcity and poverty. These lessons can turn into publicly recognized practices within local and regional spheres and can be adapted in creative forms to encourage new endogenous innovation processes.

When it comes to innovation, there are opportunities and solutions. *Opportunities* to innovate are economic, social, environmental and institutional while *solutions* (or innovations that link needs with opportunities) can be productive and service-related technologies, organizational, commercial and land use planning ones; or a combination of two or more of these solutions.

Rather than rural innovation systems, what is being seen in practice is the consolidation of cooperation networks mainly encouraged by local leaders, aimed at guaranteeing collective learning and accumulation of knowledge that comes from 'doing-using-interacting'. However these networks lack a 'linking' actor that connects research and development entities and civil society actors that *generate* knowledge and innovations, to social and productive rural organizations that *need* knowledge and innovations. This becomes a challenge that should be addressed at the local and national policy level.

There is still a need to engage in strengthening policies targeted at poverty reduction which might not in themselves be innovation policies but which can serve as bases for the fight against poverty in all its dimensions. Their conditionality in the strengthening of human capital and capabilities of the poorest constitutes a receptive environment for innovation (Fernandez-Baca *et al.* (2010).

6.4.3 Innovations in public policies at the national level

Public policies designed to fight poverty, especially rural poverty, take one of two paths:

- First, they base themselves on the 'pro-poor' orientation of the economic policies developed for agriculture and rural areas by attempting to maximize the positive externalities of these policies. The question of measuring the impact on poverty, especially on extreme poverty, still remains unanswered. This is due to the difficulty of targeting, which can lead to selection effects amongst the poor – a phenomenon we have already noted with agricultural and agrifood innovation policies.
- Second, and this is more recent, they fight poverty directly, mainly in the form of Conditional Cash Transfers (CCT). Innovative programmes have sought to improve the targeting of these policies towards the very poor and the most vulnerable. Thus, macrosocial policies for fighting poverty in rural areas (*opportunidades* in Mexico, *bolsa familia* in Brazil) have created conditionality for strengthening the human capital of the beneficiaries and have targeted, for example, women in the households as well as education for children. In this way, they have not only become significant innovations in the fight against poverty in rural areas[8] but have done so while being intrinsically disconnected from any specific reference to agriculture – which brings us back to the issue of the relationship between agriculture and rural poverty.

On their part, public policies to foster innovation in the agrifood sector mobilize the traditional actors of innovation, especially those from research and development and from extension services, and develop incentivization systems for private actors.

The importance of contextualization must be stressed, both for policies for fighting rural poverty as well as for innovation policies that are, strictly speaking, agricultural or agrifood-related. This contextualization often takes a local or regional character and may also refer to agricultural specialization. For example, in some areas where agriculture is the main source of rural household income, it is possible for policies of agricultural intensification to have positive effects on poverty. In others, however, where income sources are much more diversified, it is best to enter via rurality and a policy of territorial development. The type of agrarian structures – the relative importance of family and entrepreneurial farming and the techno-economic orientation of the production systems – are obvious key elements in this choice.

6.5 Conclusion

At the end of this review of innovation processes in rural area, we must ask ourselves the following question: Which actors and what public policies for what types of innovations? It

[8] We consider that Bolsa Familia in Brazil has been responsible for the significant drop recorded there in the Gini coefficient since the beginning of the 2000 decade.

seems relevant to us to start from the innovations and then address the issues of actors and of policies.

The range of innovations that we have had to cover is especially wide because it encompasses both agricultural innovations centred on intensification, whether 'productivist' or 'agro-ecological', as well as institutional or organizational innovations whose focus can span from the agrifood sector to entire rural areas. The sector/territory duality as it applies to the field of innovation appeared to us particularly important. Indeed, each innovation type requires the definition of 'stakeholders' with a stake in the process and of the proponents of innovation (research institutions, public organizations, businesses, etc.). While the proponents of innovation can generally be easily identified, pinpointing stakeholders is far more complex. This difficulty results from the principle that inclusion of innovation in a process of sustainable development must be based on its participatory character, which requires a compromise between stakeholders, agricultural research, farmers, farmer associations and even civil society and actors in governance system. On the other hand, for innovation to succeed the process has to be manageable at the level of the collective action of the actors involved.

As far as technological innovation in agriculture or the agrifood sector is concerned, the implementation of an innovation results in effects of selection between actors, particularly in the dissemination process. These selection effects are probably a condition for the innovation's success. As for the impact of innovation policies on rural poverty, it is probably best to include a participatory process that allows the inclusion of the 'poorest of the poor' and thus avoid the effects of selection. We are therefore faced with a dilemma which is reinforced by the gap between, on the one hand, the territorial dimension of rural development and of the reduction in rural poverty and, on the other, the sectoral dimension of technological or organizational innovations in the agrifood sector.

The concerned public policies can be divided into three heads:

- Policies targeted at poverty reduction which are not in themselves innovation policies but which can serve as bases for the fight against poverty in all its dimensions. Their conditionality in the strengthening of human capital and capabilities of the poorest constitutes a receptive environment for innovation.
- Policies for innovation in the agrifood sector, which, driven by the need for sustainable development, focus on agro-ecological intensification. Such policies are more likely than innovation policies of traditional intensification to have an explicit territorial dimension and to target family farmers. Their results can be beneficial in reducing rural poverty but this may not necessarily be their original goal.
- Rural development policies with a territorial scope. It is here that the fight against rural poverty can be harmonized with innovation for sustainable development. However, institutional innovation has to form a significant dimension of these policies.

The range of the concerned public policies therefore widely exceeds the usual definition of agricultural development policies and the commonly defined boundaries of rurality. The issue is that of a holistic and comprehensive approach of these policies, given the segmentation of skills and institutions responsible for applying them. The ability to forge relationships between them is the key factor in the contribution of such public policies to a sustainable rural development that integrates the social aspect of sustainability.

References

Boucher, F., Carimentrand, C. and Requier-Desjardins, D., 2003. Agro-industrie rurale et lutte contre la pauvreté: les systèmes agroalimentaire localisés contribuent-ils au renforcement des capacités ? 3ème Colloque sur l'Approche des Capacités, Université de Pavie, 7-9 septembre 2003.

Boucher, F., Requier-Desjardins, D. and Brun, V., 2010. SYAL: un nouvel outil pour le developpement de territoires marginaux. les leçons de l'alliance des agro-industries rurales de la Selva Lacandona, Chiapas. In: Innovation and Sustainable Development in Agriculture and Food – ISDA 2010, Montpellier, France. Available at: http://hal.archives-ouvertes.fr/hal-00521013/fr/.

Clasadonte, L., 2008. Network companies, another way on thinking agriculture. A supply chain management vision in South America. Master thesis, Wageningen University, Wageningen, the Netherlands.

Clemens, M. and Pritchett, L., 2008. Income per natural: measuring development as if people mattered more than places. Center for Global Development, Working paper no. 143, March 13, 2008.

Conway, G., 1997. The doubly green revolution: food for all in the 21st century. Penguin, London, UK.

Davezies, L., 2008. La France et ses territoires, la circulation invisible des richesses. Seuil, Paris, France.

De Grammont, H., 2010. México: boom agricola y persistencia de la pobreza rural. In: Boom agrícola y persistencia de la pobreza rural. CEPAL, Santiago, Chili, pp. 225-261.

De Janvry, A. and Sadoulet, E., 2002. Rural poverty in Latin America: tendencies and new perspectives in poverty reduction strategies. In: Pobreza Rural en America Latina y la Republica Dominicana. Mediabyte, s.a., Santo Domingo, Dominican Republic.

Dethier J.-J. and Effenberger, A., 2011. Agriculture and development: a brief review of the literature. Policy research Working Paper 5553, World Bank, Washington, DC, USA.

Dietsch, L. and Ruault, C., 2010. Dispositifs d'appui à des processus locaux d'innovation et intégration des paysans pauvres dans les montagnes sèches d'Amérique Centrale: une difficile articulation. Communication at the ISDA conference, Montpellier, France, June 2010. Available at: http://www.isda.net.

Dubois, J.L. and Mahieu, F.R., 2002. La dimension sociale du développement durable: réduction de la pauvreté ou durabilité sociale? In: J.Y. Martin (ed.) Développement durable? Doctrines, pratiques, évaluations. IRD, Paris, France, pp. 73-94.

Ellis, F. and Freeman, H., 2005. Rural livelihoods and poverty reduction policies. Routledge, London, UK.

Fernandez-Baca. E., Montoya. M.P. and Yañez, N., 2010. Innovation for poverty reduction with inclusion in the Andean region. Communication at the ISDA conference, Montpellier, France, June 2010. Available at: http://www.isda.net.

Gereffi, G., 1999. International trade and industrial upgrading in the apparel commodity chain. Journal of International Economics, 48, 37-70.

Gereffi, G. and Humphrey, J., 2005. The governance of global value chains. Review of International Political Economy, 12(1), 78-104.

Giuliani, E., 2007. The selective nature of knowledge networks in clusters: evidence from the wine industry. Journal of Economic Geography, 7, 139-168.

Griffon. M., 2006. Nourrir la planète: pour une révolution doublement verte. Odile Jacob, Paris, France.

Haggblade, S., Hazell, P. and Reardon, T., 2010. The rural non-farm economy, prospects for growth and poverty reduction. World Development, 38 (10), 1414-1441.

Hernandez, V., 2008. El fenómeno económico del boom de la soja y el empresariado innovador. Desarrollo Económico, 47 (187), 331-365.

Kline, S. and Rosenberg, N., 1986. An overview of innovation. In Landau, R. and Rosenberg, N. (eds.) The positive sum strategy. National Academy Press, Washington, DC, USA, pp. 275-305.

Lancaster, K., 1966. A new approach to consumer's theory. Journal of Political Economy, 74 (2), 132.

Lipton, M., 2006. Can small farmers survive, prosper, or be the key channel to cut mass poverty? The Electronic Journal of Agricultural and Development Economics, FAO, 3 (1), 58-85.

Nelson, R. and Winter, S., 1982. For an evolutionary theory of economic change. Harvard University Press, Cambridge, MA, USA.

North, D., 1990. Institutions, institutional change and economic performance. Cambridge University Press, Cambridge, UK.

Paillard, S., Treyer, S. and Dorin, B., 2010. Agrimonde: scénarios et défis pour nourrir le monde en 2050. Quae, versailles, France.

Parrot, L., Pedelahore, P., De Bon, H. and Kahane, R., 2010. Urban and peri-urban horticulture and the capability approach: the case of the south-west province of Cameroon. In: Coudel, E., Devautour, H., Soulard, C. and Hubert, B. (eds.). International symposium ISDA 2010. Innovation and sustainable development in agriculture and food. Available at: http://hal.archives-ouvertes.fr/hal-00516466/fr/.

Pecqueur, B., 2001. Qualité et développement rural: l'hypothèse du panier de biens et services territorialisés. Economie Rurale, 261, 37-49.

Porter, M.E., 1990. The competitive advantage of nations. Free Press, New York, NY, USA.

Rawls, J., 1971. Théorie de la justice (French translation, 1987). Le Seuil, Paris, France.

Reardon, T. and Timmer, C.P., 2007. Transformation of markets for agricultural output in developing countries since 1950. How has thinking changed? In: Evenson, R.E. and Pingali, P. (eds.) Handbook of Agricultural Economics, Vol 3: Agricultural Development: Farmers, Farm Production and Farm Markets. Elsevier Press, Amsterdam, the Netherlands, pp. 2808-2855.

Reardon, T., Barret, C. and Berdegué, J., 2009. Agrifood industry transformation and small farmers in developing countries, World Development 37 (11), 1717-1727.

Requier-Desjardins, D., 2010. International migration from Southern countries rural areas: which impact on agricultural and rural sustainability? Communication at the ISDA conference, Montpellier, France, June 2010. Available at: http://hal.archives-ouvertes.fr/hal-00521013/fr/.

Rigg, J., 2005. Land farming, livelihoods and poverty: rethinking the links in the rural south. World Development, 34 (1), 180-202.

Schumpeter, J., 1911. Théorie de l'évolution économique (French translation, 1935). Librairie Dalloz, Paris, France.

Sen, A., 1987. On ethics and economics. Basil Blackwell, Oxford, UK.

Sen, A., 1982. Poverty and famines: an essay on entitlements and deprivation. Clarendon Press, Oxford, UK.

Sen, A., 1999. Development as freedom. Oxford University Press, Oxford, UK.

World Bank, 2008. World development report 2008: agriculture for development. World Bank, Washingston, DC, USA.

Chapter 7. Quality-driven market innovations: social and equity considerations

Estelle Biénabe, Cerkia Bramley and Johann Kirsten

7.1 Introduction

What we identify as market-based innovations in agrifood systems, based in particular on contributions presented during the ISDA conference, are mainly organizational or institutional innovations involving different actors, which are linked through their direct or indirect interventions in value chains (i.e. producers, agribusiness processors, retailers, consumers, NGOs and public bodies). These links reflect increasingly sophisticated forms of intermediation and coordination in the chains (contract mechanisms and procurement schemes, certification, labelling and branding strategies, etc.) which, in many instances, represent major innovations in the chains with significant implications for sustainability. These innovations may be linked to a wide range of sustainability concerns, particularly environmentally related issues (such as resource degradation and exploitation) or unequal trade relations, and involve an increasing variety of stakeholders.

In this chapter, we focus specifically on the link between quality-driven innovations in markets and market access for smallholders, i.e. the conditions and levels of participation in food supply chains. In this regard, it is important to stress that the question of equity in market access is by no means a static one but a highly dynamic issue. This is highlighted by Berdegué *et al.* (2011) in pointing out that: '*Sustaining inclusion is much more difficult and elusive than gaining initial access to dynamic markets.*' We explore the implications of different quality-driven market-based innovations for sustainable development, building on insights from the literature and from cases that highlight the diversity of considerations and drivers for change in the chain.

7.2 Quality driven innovations and implications for small-scale farmers

7.2.1 Building on a value chain perspective

The analytical approach adopted in this chapter aims at analyzing and assessing market-based innovations and their implications from a value chain perspective. It uses the analytical tools developed mainly in the context of global value chains. As pointed out in the introduction, quality-driven innovations in markets are mainly organizational or institutional. They are based on the relations between different actors who intervene directly or indirectly in value chains. The value chain approach is particularly useful in comprehending the social implications of market-based innovations, given the conceptual frame that it provides for analyzing governance considerations in the chain and, as a result, the capacity of different actors to benefit from and sustain their participation in these chains. The governance of value chains is a function of the control of different stakeholders over the chain. It depends on the capacity of the different actors involved in the chain to dictate the rules that govern the chain (i.e. the relations between actors within the chain). In this regard, concentration at agro-processing and retail levels has favoured the domination of a few agribusiness firms in governing many agro-food chains.

Value chain governance is also framed by wider regulatory dynamics, in particular by the interaction between public or state-based regulations and private regulatory activities. The manifestations of these dynamics can be peculiar to specific chains. As pointed out during the discussions at the Conference, the consideration of governance issues and the interaction between state-based regulation and private stakeholders' actions help bridge the conceptual divide, which is often too widely assumed, between the market and the State.

7.2.2 Quality-driven innovations and the sophistication of coordination in value chains

As evidenced at the Conference, quality-driven innovations in markets are mainly associated with the development of standards. It is now widely recognized that trends towards quality-oriented supply chains that operate through standards are significantly modifying the modes of coordination within these chains. They are reshaping the organization of production and trade relations, thereby affecting governance and market participation. Indeed, many authors have highlighted, in global and quality-oriented food supply chains, the move away from open spot markets with anonymous suppliers and lack of proper accountability towards higher degrees of vertical coordination (see, among others, Gibbon and Ponte, 2005). According to Ruben *et al.* (2006), this increased degree of vertical coordination based on complex contractual arrangements occurs together with the increased monitoring of product quality and process standards. It is accompanying the move from competition between individual actors to competition between supply chains (Hanf and Pienadz, 2007). Strengthened coordination among actors along supply chains is a key factor in meeting new quality dimensions and therefore ensuring differentiation. This is in line with the empirical observations and analysis by Wilson *et al.* (2000) of the market performance of two protected denomination of origin products (i.e. early potatoes from the United Kingdom and from the Netherlands). These authors show that different levels of cooperation and coordination in the chain result in significant differences in product specification, brand promotion and consumer awareness.

It is conversely important to point out that, even though strengthened vertical coordination and more sophisticated quality management systems are used as part of differentiation strategies by lead operators in the chain, collective action has also been emerging between large supply chains and companies, in particular at the retail level. The GlobalGap standard is one particularly significant illustration of retailers' attempt at standardizing international supply chains (Henson and Reardon, 2005). The move from retailer-specific standards to standards that regulate a group of retailers and which are monitored through third party certification triggers a tightening of horizontal coordination together with the adoption of these private collective voluntary standards (Havinga, 2008). It arises from the common interests in limiting the transaction costs for setting up standards. Fulponi (2006a) investigates the emergence of coalitions of firms for setting private collective voluntary standards as a key development in food systems. In addition to GlobalGap, examples of this include the Eco-friendly standard, the IFOAM organic guidelines, the Fair Trade Initiative and the Ethical Trading Initiative. Hammoudi *et al.* (2009) confirm that these standards entail both vertical and horizontal coordination.

Studies based on transaction cost economics help deepen the understanding of the move towards more coordinated supply chains together with more complex quality management systems. Raynaud *et al.* (2002) analyzed governance structures in 42 case studies, conducted in three different agrifood sectors and in seven European countries. They show that the agent developing a quality signal, and whose value depends on other agents in the chain, specifically designs mechanisms for governing transactions at bilateral level so as to ensure credibility through a proper product guarantee scheme. This may result in different types of contractual relations, depending on the quality signals. Barcala *et al.* (2007) also stress that different governance mechanisms and organizational forms have varying impacts on product quality. Their results indicate that the most market-oriented governance mechanisms in their study (quasi-integrations and geographical indications) go together with (1) coordination mechanisms such as norms and routines to perfectly define standards and attributes and (2) a complementary set of quality control devices based on direct supervision. In the same vein, Ménard and Valceschini (2005) point out that the type of quality signal influences the governance of transactions, stating that private brands are more often associated with vertical integration than brands that are linked to public certification[1]. Ménard and Valceschini (2005) further highlight the importance of considering the link between supply chain-based governance structures that are designed to address specific quality signaling issues and the macro level at which institutional mechanisms are developed in support of the credibility of the adopted modes of organization (e.g. increased use of third party certification accredited by public institutions).

[1] Certification is a procedure by which a third party gives written assurance that a product, process or service is in conformity with certain standards (ISO Guide, 1996). The certifying organization is called the certifier or certification body. The certifier may conduct the audit/inspection itself or contract it out to an auditor/inspecting body. The system of rules, procedures and management for carrying out certification, including the standard against which a company is certified is called the certification programme. One certification body may execute different certification programmes. To ensure that certification bodies have the capacity to carry out the certification programme, they are evaluated and accredited by an authoritative institution (FAO, 2007).

Interestingly, Ponte and Gibbon (2005) argue that the capacity to capture complex information over quality in standards, labels, certification and codification, lowers the need for vertical integration which arises from increased quality signaling and management complexity. Indeed, in many high-value supply chains, lead firms employ standards and branding strategies to exercise control over suppliers without necessarily establishing ownership structures (UNCTAD, 2007). Ponte and Gibbon (2005) emphasize the role played by the control over the qualification mode and information management in the capacity of lead firms to exercise their 'functional leadership'. They show that firms' capacity to transfer relatively intangible information to their suppliers, to standardize and/or to obtain credible external certification for increasingly complex quality content of goods and services, allow for relatively loose forms of coordination and high level of drivenness. This is supported by UNCTAD (2007) which stresses the increased importance of controlling and owning intangible assets, in particular information and brands, rather than controlling the physical means of production, as ways of dealing with competition and governing supply chains.

7.2.3 The need for exploring implications for small-scale farmers of different quality related supply chain dynamics

Reardon *et al.* (2003), among others, have pointed out the effect of more sophisticated supply chains on the reinforcement of downstream players' bargaining power and their implications in terms of the exclusion of small-scale farmers from food supply chains, particularly in the context of developing countries. Codron *et al.* (2005) discuss these issues in the context of quality strategies adopted by retailers when faced with a mandatory quality management system. They stress that the vertical relationship between retailers and suppliers and, therefore, the distribution of bargaining power between them, determines the allocation of costs associated with implementing the quality strategy. Vorley (2001) discusses these issues in the context of standards development and requirements related to sustainability considerations developed in response to pressure from civil society (NGOs). He argues, in line with Ponte and Gibbon (2005), that these standards contribute to reinforcing the control of major downstream players on the supply chain and to increasing barriers to market entry. Indeed, downstream players therefore have the capacity to shift the burden of compliance costs and risks to their suppliers. This allows downstream players to actively engage in farm level decisions without vertical integration. A good example of this type of retailers' intervention at the farm level is the GlobalGAP standard which requires Good Agricultural Practices (GAPs) in primary production and extends the principles of risk identification and management to farm production. The decrease in bargaining power of primary producers is particularly concerning in light of the withdrawal of State support from agriculture and market intervention.

The development of vertically coordinated supply chains, characterized by the use of private standards and controlled by major role players in the agrifood industry, has changed the rules for market participation. It has resulted in an increased disjunction between the price at producer level and that at the market price (Vorley, 2001). In many parts of the world, this contributes significantly to confining small-scale farmers to local low-income markets (Ssemwanga, 2005). It marginalizes small-scale farmers from high-value supply chains (including export markets) and prevents them from benefiting from quality trends. However, Vorley (2001) also points out the process of rural differentiation and of diversification currently taking place in the smallholder economy and the need to understand its implications. He argues that, in the widely shared context of a liberalized environment, market access depends on the capacity to exploit 'marketing advantage'. He therefore stresses the issue of better understanding the actual changes in the terms of trade between producers and downstream role players in the supply chains in different situations. There is a need to identify the nature of the upgrading in different supply chains to relate it to the innovation capacity of the actors and to assess its implications. As noted by Giovannucci (2003), social and environmental attributes are moving from differentiating factors into mainstream market criteria to increasingly becoming necessary conditions for inclusion in the more developed markets, thereby more strongly affecting small-scale farmers. It is thus extremely important to understand the capacity and limitations of market innovations and small-scale farmers for developing and benefitting from 'marketing advantage'. The following section intends to provide insights into this by drawing on both the literature and the discussions at the Conference on a number of market-based innovations in different countries.

7.3 Mixed evidence on the implications for small-scale farmers of quality developments in the chains

7.3.1 General insights into the exclusionary effects of quality developments

As indicated in the previous section, quality-based supply chains and the increasing use of private food standards clearly have implications for the participation of small-scale farmers. Indeed, it has been argued that private standards are fast becoming the key factor in determining market access (Henson and Reardon, 2005). Compliance with standards implies significant capacities in accessing the relevant information, in adapting production practices and implementing new processes (Giovannucci and Reardon, 2000). Furthermore, Chemnitz et al.(2007) shows that private standards are generally more costly than government requirements with respect to communication and documentation of the certification process and may, therefore, hold more serious exclusionary implications. These exclusionary implications are exacerbated by the fact that private retailer standards, in addition to traditional product controls, focus on management processes to achieve a given outcome (Fulponi, 2006a). The lack of harmonization across public and private standards further complicates standard-based requirements and therefore poses an increasing challenge to the participation of producers in supply chains (see, for example, Henson and Humphrey, 2009).

Farina *et al.* (2005) point out that there is evidence, in terms of investment and management, that this will more seriously affect small-scale farmers, particularly in developing countries.

Economies of scale is an important factor that negatively impacts the capacity of small-scale farmers to adhere to an increasingly stringent standards environment and the resulting higher cost of compliance (Dolan and Humphrey, 2000). This aggravates the already unequal capacities of small-scale and large-scale producers to supply lead firms which demand high levels of production and, in many instances, prefer to cooperate with larger farmers due to the high communication and monitoring costs when dealing with small-scale producers (Swinnen, 2005). Vorley (2001) also questions the ability of small-scale producers to exploit 'marketing advantage' when the market favours the capacity of large-scale farmers to handle post-harvest processes and transport over small-scale producers' provision of higher quality at a lower cost. Small-scale producers may have some competitive advantage in a number of quality-based markets (e.g. organics) in terms of labour intensive processes, etc. However, the above-described conditions for market participation in vertically coordinated supply chains dominated by private standards often prevent small-scale producers from capitalizing on their 'marketing advantage'.

However, as stressed by Swinnen (2005), observations are inconclusive regarding the level of exclusion of small-scale farmers. While most studies point to exclusionary dynamics, there are also a number of studies that allude to the potential for inclusion of small-scale producers in quality-based supply chains in developing countries (Giovannucci, 2003; Ponte and Gibbon, 2005). The following sub-sections further explore and discuss options for inclusion arising from quality-based market innovations.

7.3.2 Insights into the potential for inclusion of contractual arrangements in quality-based chains: knowledge considerations

The increasing importance of standards and coordination mechanisms in the chains has been found to be not only a barrier to entry but also an opportunity for small-scale farmers to engage in a learning process that could improve their access to markets. Indeed, standard specifications present knowledge in a codified and packaged form (Fulponi, 2006b). Furthermore, information management on product quality may be eased by the use of standards, therefore making it less tedious to procure from a large number of small-scale producers (Chemnitz et al., 2007). By strengthening relationships between actors, coordinated chains have the potential to enhance the capacity of small-scale farmers to understand, and therefore to comply with, buyer requirements and changing needs.

In this regard, the conference paper presented by Bolo (2010) on the cut-flower industry in Kenya provides interesting insights (see Box 7.1). It illustrates some limits in the role played by the establishment of partnerships and contractual relationships between small growers and exporters in building farmer capacities. Although this arrangement has provided market access and therefore contributed to overcoming the exclusion of smallholders from an important high-value export supply chain, it is being argued that the relationship is one-dimensional, with growers almost only involved in production activities under strict

Box 7.1. Learning to export: building farmer capabilities through partnerships in Kenya.
Maurice Ochieng Bolo

Contractual partnerships between smallholder farmers and exporters are viewed as necessary for building the capability of smallholders, while affording the exporters oversight to ensure timely delivery of high-quality products. This study surveyed 116 farmers/exporters using a structured questionnaire. It then focused on three case studies of on-going contractual partnerships in order to provide a detailed description of how the institutions and governance arrangements influence farmer capabilities (Bolo, 2010). Our results demonstrate that in terms of learning, farmers rated production capabilities at 58.6% followed by marketing capabilities at 23.3%. Value-addition capabilities were rated lowest, at 18.2%. This result is attributed to the institutional arrangements crafted in the contracts, as discussed below.

Contracts as institutions

These 'farmer/exporter' partnerships are prone to a number of challenges. This includes exporters having to deal with a large group of weak suppliers, thus increasing their transaction and coordination costs. The exporters face the risk of opportunistic behaviour from the farmers, leading to the possibility that competitors could tap into the pool of skilled suppliers, which were developed by another exporter. These challenges keep the exporters constantly under pressure, hence the need for them to ensure protection measures are in place. This is achieved through institutions – the rules, laws, norms and codes that regulate behaviour – as embodied in the contracts.

Achieving 'lock in – lock out'

Clauses in the actual contracts show that exporters are required to provide training to farmers regarding the production of 'good quality flowers'. This training includes activities ranging from planting to harvesting but excludes post-harvest activities, which are reserved for the exporters. Similarly, clauses that relate to value-adding activities, including grading, bunching and packaging, are assigned to the exporters. Farmers only carry out preliminary grading (sorting) under the supervision of the exporters. Further value-addition is conducted at the exporters' pack-house where flowers are sleeved and subjected to pre-treatment solutions before they are packaged and wrapped for shipment. In order to lock-in farmers, the contracts prohibit farmers from entering into contracts with other exporters to supply the same crops.

Post-harvest handling, value addition and customer specifications form the 'success-limiting steps' for smallholders. They constitute 'specialized' knowledge held by the exporters who also own the technical infrastructure for value addition. Since learning is incremental (Cohen Wesley and Levinthal, 1990), and increased capabilities are a source of competitiveness (Teece et al., 1998), reduced costs, improved performance and reliability (Levinthal and March, 1993), the knowledge gained from these partnerships forms the 'receptor sites' for advanced knowledge in the future. However, smallholders may fail to develop 'receptor sites' for value-adding knowledge and end up locked into production – thereby increasing their marginalization (Bolo, 2010).

guidelines and instructions. The procurement contracts provide for training and the exporters assist in building farmer capacity to meet the requirements. However, training is predominantly limited to production-related activities, with exporters generally not sharing knowledge attached to value-adding activities (post-harvest handling and selling activities). This excludes farmers from participating in these activities.

Other contributions at the conference also highlighted the incomplete learning process arising from the contractual relationships between smallholder farmers and downstream role players, even in cases where the downstream operator is more committed to empowering farmers through sustained contractual relationships (see Dulcire, 2010 on cocoa in Sao Tome). Dulcire (2010) questions the capacity of small-scale farmers to build production and marketing skills and to become less dependent on external intervention. He pointed out that, even though downstream operators are interested in securing quality supply from small-scale farmers as part of a quality labelling strategy (organic and fair trade), the fact that the terms of the contract are mostly externally determined and imposed, strongly affects the farmers' ability to take ownership of the process, resulting in incomplete learning both technically and organizationally.

7.3.3 Insights into the potential for inclusion of changes in the trading rules

As pointed out by Altenburg (2006), changes in the rules and conditions for supplying chains that have been developing quality strategies may in themselves favour small-scale producers.

Fair trade rules

Loconto and Simbua (2010) investigated the implications of fair trade rules in tea value chains in Tanzania. Their results, while not confirming Altenburg's statement, are of interest (see Box 7.2), as fair trade rules are meant to change the rules for market participation to the benefit of small-scale farmers. Loconto and Simbua question the actual level of influence of fair trade rules on the governance in tea value chains in Africa. They show that the specific and prior organizational arrangements in the chains between smallholders and factories are more determinant on the capacity to penetrate and sustain fair trade market participation for smallholders and in structuring the relations in the chains than on the characteristics of the fair trade chains that would arise from the principles and governing rules underlying fair trade certification (price premium and investment in development, advance payment possibility, long-term engagement through contract). Long-term stakeholder involvement contributes towards more balanced and sustainable relationships between growers and processors in the tea sector. Ownership arrangements are not determinative of sustained fair trade participation. The capacity to sustain participation in fair trade chains therefore appears to be more a result of balanced relations than of drivers for developing equity and the participation of small-scale farmers in the chains.

Box 7.2. *Fair trade standards as a means to innovate in the organization of the chain?*
Alison Loconto and Emmanuel Simbua

This case questions the influence of fair trade rules on the governance of value chains in Africa, specifically on the Fairtrade objective of small-farmer empowerment. Using data from the tea sector in Tanzania, Loconto and Simbua explored the existing organizational arrangements between smallholder producers and tea processing factories that are part of Fairtrade certified value chains. Their results show that a diversity of ownership arrangements exists in the Fairtrade certified value chains and that these arrangements function rather independently of the Fairtrade certification system. Indeed, historical and institutionalized relationships between factory management and smallholders either facilitate or hinder the involvement of producers in Fairtrade markets.

Loconto and Simbua (2010) note in particular that the tea processing factories that are part of the Fairtrade network are those that have been moving towards organizational innovation as part of their more general business models. This innovation includes joint-ownership of tea processing factories by smallholder cooperatives and private investors, and collectively negotiated outgrower contracts for each tea processing factory. These outgrower contracts consist of input subsidies and credit for the farmers and an assurance of on-time payment on the part of the factories. Moreover, the locally negotiated rates for payment are highly influenced by local power relations, where smallholder cooperatives carry more influence than is commonly assumed. Finally, the authors found that only one out of the four case studies consisted of a smallholder association that has been able to successfully maintain its Fairtrade certification.

The authors suggest that organizational innovation, which effectively includes increased stakeholder involvement in decisions and control over resources, is important for success in maintaining Fairtrade certification. However, they also found that the perceived benefits from Fairtrade do not necessarily increase with ownership shares in the processing factory; rather it is the long-term relationships between value chain actors and other stakeholders that have worked to create more sustainable relationships between growers and processors in the tea sector. Hence those characteristics of Fairtrade value chains (e.g. price premium and investment in development, advance payment possibility and long-term engagement through contracts) are, for the most part, already present in some of the tea value chains in Tanzania. In sum, it would appear that it is these existing conditions that facilitate participation in Fairtrade value chains, rather than *vice versa* (Loconto and Simbua, 2010).

Larry and Stevenson (2010) conversely shows the shared interest of mid-sized farms and agro-food enterprises, operating at regional level in the United States, in building unique 'fair trade' business models. These models consist of strategic alliances based on equity considerations, transparency and shared brand identities (unique food stories) that foster consumer trust and support. In these instances, specific 'fair trade' rules are developed internally between farmers and agro-food enterprises as part of developing specific quality angles for differentiating their products and increasing their competitiveness in regional markets. While not focused on small-scale farmers, these models allow farmers that are neither competitive in international markets nor adapted to direct marketing at local level, to maintain their activities and thereby contribute to social (community vitality) and environmental (diverse and resilient structure of agriculture) sustainability. These models rely on collective action both at the horizontal level among farmers and at the vertical level between these farmers and the downstream enterprise.

Certification processes

Contributions to the conference also provided insights into the role of certification processes and rules in changing and accounting for social dynamics associated with markets, and therefore their potential for including small-scale farmers. Sabourin (2010) showed that certification schemes involving collective processes (co-certification, group certification and participative guarantee schemes) can play a role in reintroducing reciprocity in the 'capitalist exchange market system' and therefore support different market-based social dynamics. He shows in particular that these types of schemes offer better opportunities for accounting for small-scale farmers' specific characteristics and for valuing them.

Other contributions to the conference also supported the importance of questioning the capacity of different standards and quality-based mechanisms to actually reflect, take advantage of and value the diversity of production systems and conditions. Mateos and Ghezán (2010) emphasize the importance of considering these issues in national organic standards regulatory processes by contrasting the Argentinean and Brazilian cases (see Box 7.3). They analyze the Argentinean regulatory trajectory and the changes in considerations brought about by the evolution of social dynamics. Their analysis of the Argentinean organic standards regulations raises questions as to which stakeholders contribute to policy-making and to the nature of the public regulation that shapes the activities of the actors and the functioning of the market. This again stresses the importance of not considering the State and the market as opposing forces but as closely intertwined dimensions.

7.4 Conclusion

The focus of this chapter has been on market-based or chain innovations in agrifood systems. These innovations have grown substantially in recent years, together with the dynamics of quality and the development of standards. They are strongly associated with increasingly sophisticated forms of intermediation and coordination in the chains. This impacts the governance of chains and presents significant implications for sustainability from a social perspective. As shown in this chapter, the potential of market innovations to drive changes towards increased and sustained inclusion of small-scale farmers strongly depends on who

Box 7.3. Social construction of quality standards in organic agrifood and the inclusion of small-scale producers. Insights into the Argentinean case.
Mónica Mateos and Graciela Ghezán

In the case of organic products, the analysis of the social construction of quality processes allows for an understanding of the co-existence of multiple conventions (Busch, 2004). The commercial norms, developed through hybrid forums (Callon et al., 2002) or socio-technical networks, highlight the controversies over the inclusion of small-scale producers.

The Argentinean case illustrates the predominance of market considerations over those regarding the inclusion of small-scale farmers in the organic standards regulation. The majority of Argentinean organic food production (95%) is sold on the international market, with Argentina being the first third-party country to be certified organic by the EU. During the 1990s, the legal requirement for third-party certification did not consider the difficulties of small-scale producers in adapting production practices and implementing new processes.

Interestingly, in Latin America, the evolution of organic food regulations have changed over the years. Initially the Brazilian organic dynamics were the exact opposite of the Argentinean dynamics, with social movements and non-governmental organizations (NGOs) attempting, from the start, to include small-scale producers and accommodate the concerns of local consumers by developing locally adapted ecological agriculture. In this regard, Brazil does not only account for the market considerations imposed by international markets (Meirelles, 2003). The recent transformations are principally related to the different social, economic and political contexts. In recent years, there have been increased information and knowledge about various experiences that include small-scale producers. The heterogeneous cases of inclusion go from fair trade to the development of ecological agriculture.

Actually in Argentina, as evident from the analysis of the negotiations on the different dimensions of quality in organic products, the recent promotion of family agriculture, based on sustainability and the diffusion of collective certification processes, opens a new debate about the norms and rules implemented until now.

Hence in the process of change of the law on organic production, the purpose is to include small producers from different perspectives such as: differentiate the type of certification required according to the markets (domestic or international) and the inclusion of new forms of participative certification (through social movements/producer associations) with low or high participation from the State. The direction and speediness of the changes are related to the capacity of these actors to link up with national/regional innovative networks on productive as well as on organizational aspects (Mateos and Ghezán, 2010).

controls or drives these innovations. Structural asymmetric power is still very prevalent and hugely constrains the capacity for market innovations to improve the participation of small-scale farmers. In many instances, quality and market innovations have been shown to reinforce dominant positions. Market innovations may valorise local practices and resources as marketing advantages and could thereby potentially enhance market access for small-scale farmers. However, this critically depends on the nature of the processes and, in particular, on the possibility for more balanced relationships. Different dimensions were shown to be important in this regard. These include the potential for and nature of the learning and empowerment processes, with knowledge considerations encompassing both technical and organizational dimensions. Also included is the nature of the regulatory processes, as relating to the development of certifications and standards. Furthermore, aspects of collective action appeared to be a critical underlying dimension in the different cases analyzed. In addition to the market considerations developed in this chapter, many of the contributions presented during the ISDA conference also highlighted the importance of collective action around quality building and signalling at the local level to take advantage of specific resources (e.g. territorial assets) through the co-construction of territorialized/local resources connected with quality and know-how.

All these three dimensions – the collective-action dimension and the social dynamics behind the development of standards and processes of qualification, the dimension of knowledge distribution and the rules attached to the regulatory processes – need to be critically scrutinized when considering the potential of different market innovations to enhance the inclusion of small-scale farmers. They simultaneously determine and are influenced by the nature and scope of the markets in which these chain-based innovations take place. This includes, in particular, the geographical scale of the market (local, regional, international) and the type of product attributes that are promoted in the chain (fair trade, organic, etc.).

References

Altenburg, T., 2006. Introduction to the special issue: shaping value chains for development. The European Journal of Development Research, 18, 493-497.

Barcala, M.F., Gonzalez-Diaz, M. and Raynaud E., 2007. The governance of quality: the case of the agrifood brand names. In: 3rd International Conference on economics and management networks. Rotterdam School of Management. Erasmus University, 28-30 June 2007.

Berdegué, J., Biénabe, E. and Peppelenbos, L., 2011. Conclusions: innovative practices in connecting small-scale producers with dynamic markets. In: Biénabe, E., Berdegué, J., Peppelenbos, L. and Belt, J. (eds.). Reconnecting markets: innovative global practices in connecting small-scale producers with dynamic food markets. Gower Publishing, Farnham, UK.

Bolo, M.O., 2010. Learning to export: building farmers' capabilities through partnerships in kenya's flower industry. In: Innovation and sustainable development in agriculture and food – ISDA 2010, Montpellier, France. Available at: http://hal.archives-ouvertes.fr/hal-00526145/fr/.

Busch, L., 2004.Grades and standards in the social construction of safe food: In: Lien, B. and Nerlich, M. (eds.) The politics of food. Berg Publishers, Oxford, UK, pp. 163-178.

Callon, M., Meadel, C. and Rabeharisoa, V., 2002. The economy of qualities. Economy and Society, 31(2), 194-217.

Chemnitz, C., Grethe, H. and Kleinwechter, U., 2007. Quality standards for food products – a particular burden for small producers in developing countries? Contributed paper at the EAAE Seminar 'Pro-poor development in low income countries: food, agriculture, trade and environment', 25-27 October 2007, Montpellier, France.

Cohen Wesley, M. and Levinthal, D.A., 1990. Absorptive capacity: a new perspective on learning and innovation. Administrative Science Quarterly, 35 (1), 128-152.

Codron, J., Giraud-Héraud, E. and Solar, L., 2005. Minimum quality standards, premium private labels, and European meat and fresh produce retailing. Food Policy, 30, 270-283.

Dolan, C. and Humphrey, J., 2000. Governance and trade in fresh vegetables: the impact of UK supermarkets on African horticultural industries. Journal of Development Studies 35, 147-177.

Dulcire, M., 2010. La mise en place participative d'une filière cacao à Sao Tome. L'organisation des producteurs en tant que facteur d'émancipation. Communication presented at the Innovation and Sustainable Development in Agriculture and Food (ISDA) conference, Montpellier, France, June 28-30, 2010. Available at: http://hal.archives-ouvertes.fr/docs/00/51/05/55/PDF/Dulcire_cacao_Sao_TomA_.pdf.

Food and Agriculture Organization (FAO), 2007. Private standards in the United States and European Union markets for fruit and vegetables – implications for developing countries. FAO commodity studies. Available at: www.fao.org.

Farina, E.M.M.Q., Gutman, G.E., Lavarello, P.J., Nunes, J. and Reardon, T., 2005. Private and public milk standards in Argentina and Brazil. Food Policy, 10, 302-315.

Fulponi, L., 2006a. Private standards and the shaping of the agro-food system. OECD report, Paris, France.

Fulponi, L., 2006b. Private voluntary standards in the food system: the perspective of major food retailers in OECD countries. Food Policy, 31, 1-13.

Gibbon, P. and Ponte. S., 2005. Trading down: Africa, value chains and the global economy. Temple University Press, Philadelphia, PA, USA.

Giovannucci, D., 2003. Emerging issues in the marketing and trade of organic products. Proceedings of the OECD Workshop on Organic Agriculture, September 2002. OECD, Paris, France.

Giovannucci, D. and Reardon, T., 2000. Understanding grades and standards and how to apply them. World Bank, Washington, DC, USA.

Hammoudi, A., Hoffmann, R. and Surry, Y., 2009. Food safety standards and agri-food supply chains: an introductory overview. European Review of Agricultural Economics, 36 (4), 469-478.

Hanf, J.H. and Pienadz, A., 2007. Quality management in supply chain networks. The cases of Poland. International Food and Agribusiness Management Review, 10 (4), 103-128.

Havinga, T., 2008. Actors in private food regulation: taking responsibility or passing the buck to someone else? Paper presented at the symposium 'Private Governance in the Global Agro-Food System' Munster, Germany, 23-25 April 2008.

Henson, S. and Reardon, T., 2005. Private agri-food standards: implication for food policy and the agri-food system. Food Policy, 30, 241-253.

Henson, S. and Humphrey, J., 2009. The impacts of private food safety standards on the food chain and on public standard-setting processes. Joint FAO/WHO Food Standards Programme, Codex Alimentarius Commission, Thirty-second session, 29 June – 4 July 2009, Rome, Italy.

Larry, L. and Stevenson, J., 2010. Acting collectively to develop midscale food value chains in the U.S. Communication presented at the Innovation and Sustainable Development in Agriculture and Food (ISDA) conference, June 28-30, Montpellier, France. Available at: http://hal.archives-ouvertes.fr/docs/00/52/04/62/PDF/LEV_acting_collectively.pdf.

Levinthal, D.A. and March, J.G., 1993. The myopia of learning. Strategic Management Journal, 14, 95-112

Loconto, A.M. and Simbua, E.F., 2010. Organizing smallholder production for sustainability lessons learned from fairtrade certification in the Tanzanian tea industry. In: Innovation and Sustainable Development in Agriculture and Food – ISDA 2010, Montpellier, France. Available at: http://hal.archives-ouvertes.fr/hal-00529061/fr/.

Mateos, M. and Ghezán, G., 2010. El proceso de construcción social de normas de calidad en alimentos organicos y la inclusión de pequeños productores el caso de Argentina. In: Innovation and Sustainable Development in Agriculture and Food – ISDA 2010, Montpellier, France. Available at: http://hal.archives-ouvertes.fr/hal-00566243/fr/.

Meirelles, L., 2003. La certificación de productos orgánico-caminos y descaminos. Available at: http://www.centroecologico.org.br/artigo_detalhe.php?id_artigo=25.

Ménard, C. and Valceschini, E., 2005. New institutions for governing the agri-food industry. European Review of Agricultural Economics, 32 (3), 421-440.

Ponte, S. and Gibbon, P., 2005. Quality standards, conventions and the governance of global value chains. Economy and Society, 34(1), 1-31.

Raynaud, E., Savée, L. and Valceschini, E., 2002. Governance of the agri-food chains as a vector of credibility for quality signalization in Europe. In: 10th EAAE Congress, 'Exploring diversity in the European Agri-food System', August 28-31, Zaragoza, Spain.

Reardon, T., Timmer, C.P., Barrett, C.B. and Berdegue, J., 2003. The rise of supermarkets in Africa, Asia, and Latin America. American Journal of Agricultural Economics, 85 (5), 1140-1146.

Ruben, R., Slingerland, M.A. and Nijhoff, H., 2006. Agro-food chains and networks for development: issues, approaches and strategies. In: Ruben, R., Slingerland, M.A. and Nijhoff, H. (eds.) Agro-food supply chains and networks for development. Wageningen UR-Frontis Series 14, Kluwer, Dordrecht, the Netherlands, pp. 1-25.

Sabourin, E., 2010. Agri-food qualification and certification process as an interface between exchange marketing and reciprocity. Communication presented at the Innovation and Sustainable Development in Agriculture and Food (ISDA) conference, Montpellier, June 28-30, France. Available at http://hal.inria.fr/docs/00/52/19/69/PDF/Sabourin_Agri-food.pdf.

Ssemwanga, J., 2005. Presentation at the final meeting of the EU concerted action 'Safe and high quality food chains'. Buenos Aires, 22 May 2005.

Swinnen, J.F.M., 2005. When the market comes to you – or not: the dynamics of vertical coordination in agri-food chains in transition. In: Final Report of the World Bank (ECSSD) ESW on Dynamics of Vertical Coordination in ECA Agrifood Chains: Implications for Policy and Bank Operations. World Bank, Washington, DC, USA.

Teece, D.J., Pisano, G. and Shuen, A., 1997. Dynamic capabilities and strategic management. Strategic Management Journal 18, 509-533.

United Nations Conference on Trade and Development (UNCTAD), 2007. Private-sector-set standards and developing counties' exports of fresh fruit and vegetables: synthesis of country-case studies in Africa (Ghana, Kenya, Uganda), Asia (Malaysia, Thailand, Vietnam), and Latin America (Argentina, Brazil, Costa Rica). Background note by the UNCTAD secretariat for the FAO-UNCTAD Regional Workshop on Good Agricultural Practices in Eastern and Southern Africa: Practices and Policies, Nairobi, Kenya, 6-9 March 2007.

Vorley, B., 2001. The chains of agriculture: sustainability and the restructuring of agri-food markets. IIED and RING opinion paper prepared for the World Summit on Sustainable Development. Available at: www.ring-alliance.org/ring_pdf/bp_foodag_ftxt.pdf.

Wilson, N., Van Ittersum, K. and Fearne, A., 2000. Cooperation and coordination in the supply chain: a comparison between the Jersey Royal and the Opperdoezer Ronde potato. In: Sylvander, B., Barjolle, D. and Arfini, F. (eds.) The socio-economics of origin labelled products in agro-food supply chains: spatial, institutional and co-ordination aspects. Series Actes et Communications, 17 (1). INRA, Paris, France.

Chapter 8. Innovation and governance of rural territories

André Torre and Frédéric Wallet

8.1 Introduction

The idea that innovation or creativity can be the basis of the processes of development of territories has appeared only fairly recently in the literature and in public policies and actions. And it is only in the past few years that there has been an acceptance of the fact that new activities can be useful to – or even be drivers of – the growth of rural territories (Regional Science Policy and Practice, 2011). And yet, this approach is still usually confined to the high-technology or new economy sectors.

It was only in the 1990s that work was undertaken that placed innovation at the core of regional or territorial growth. It highlighted the importance, in this mechanism, of innovative firms and of clusters that brought together high-tech creative activities. It discussed the spatial dissemination of technologies and its geographic limits in terms of spillovers. Also noted were the problems relating to the capacity of absorption and the difficulties of reproducibility of innovations developed elsewhere. This movement resulted in, and was accompanied by, the implementation of many local, national and community policies according priority to innovation, such as the creation of science parks and technopoles, the significantly increased R&D funding or the strengthening of research-industry relationships. Almost without exception, it was the development of high-tech innovations that was favoured, with an emphasis on the creation and transfer of innovations of a very high level. They were supposed to benefit the enterprises that used them as well as the network of their subcontractors, suppliers or geographical neighbours and, through a trickle-down effect, the entire local economy.

The resulting model of regional or territorial development is therefore based on high-tech activities. Innovation is considered the main engine of growth (a watered-down version of development) as well as a differentiating factor useful for overcoming competitive constraints, at least partially. International institutions (OECD, EU, etc.) and national governments advocate these development policies based on innovation and competitiveness and have set up mechanisms to intensify selection between territories. This often results in land planners and managers acquiring a naive and wishful attitude, wanting to enter a competitive world and considering that valorising local resources and supporting cutting-edge sectors are enough to generate development.

But the territories are not at an equal footing in the race for technological excellence since not all have resources that can easily be valorised or the expertise necessary to do so. This is especially true for rural territories – and for countries of the South – and thus the question of the nature of innovation and the conditions under which it can truly bloom in territories needs to be readdressed. This chapters's goal is to explore the links between three key elements: innovation, territorial development and governance. In the first part, we present the main development models and the various types of their implementations in rural or agricultural territories. We then discuss the role of innovation in development approaches by considering successively the approaches of territorialized innovation and policies of territorial innovation. We conclude with an analysis of modes of governance of rural and periurban spaces as expressions or vectors of innovations in territories. This analysis covers processes of negotiation and decision making, actors and governance structures and mechanisms dealing with conflicts or encouraging consultation.

8.2 Models of regional and territorial development

Work on the theme of development, whether focused on rural and agricultural issues or more generally attempting to define conditions for the growth and success of regional economies, most often take the form of studies of economic mechanisms. It is readily apparent that the issue of innovation, of limited interest during the post-war boom years, has now become a major component of these approaches, given that development is now closely linked to innovation in all its forms. Three major competing visions of development currently coexist, corresponding to strong analytical assumptions in which innovation is present to a lesser or greater degree (Torre and Wallet, 2013).

8.2.1 Development as an optimum balance

First of all is the thinking that focuses primarily on defining a balance between interests and gains derived by the various local actors of the development process and on seeking principles that will lead to the maximum satisfaction of all stakeholders. The founding approaches of neoclassical theory belong to this category. They propose a homothetic growth based on inputs of capital and work, later extended to a third, more technological, input, most often in the form of knowledge or the amount of R&D investments (Solow, 2000). In these approaches, innovation is mainly considered as an input that can improve the efficiency of the allocation and use of production factors and thus boost productivity. It is a matter of assessing the production volume and its growth and of comparing them to the optimal combination of factors and the efforts undertaken in terms of productivity or capital accumulation for example (Johansson *et al.*, 2001). This approach, which envisages the eventual possibility of eliminating interregional disparities, has seen significant success and has only been held back by its limitations in terms of homothetic growth and of its inability to account for imbalances reported early on by the authors of polarization or by growth at the bottom, for example.

8.2.2 Development as a source of inequality and polarization

The second, and largest, group of analyses consists of approaches that consider compromises made between local actors to be only temporary and ultimately untenable. They believe development processes always generate interregional inequalities which are hard to reduce. In contrast to the 'optimum balance' thinking, these analyses consider that development brings and contributes to the widening of disparities between regions or territories, often permanently. They also highlight the existence of local systems with specific institutional, economic or technical characteristics and whose successes or failures induce fundamentally unbalanced development processes. This body of work is based on the analyses of growth poles, conceived by Perroux and developed by Mydral, Hirschmann and Higgins. Perroux's (1961) original idea is that development cannot occur everywhere at the same time and with the same intensity. This is amply demonstrated by countries or areas that are lagging behind in development, a fact that the growth pole theory was the first to recognize. Development relies on a process of polarization of activities, itself based on the existence of large companies which act as driving forces, located in the heart of the most developed regions. They are the vectors of innovation and of its unbalanced dissemination between territories.

With the crisis of Fordism and the inability of traditional models to account for changes in capitalism, such as the success of forms of organizations other than the large-company model, new analyses have emerged which place intangible factors at the heart of the dynamics of development. Thus, Porter (1985, 1990), whose approaches have had a wide impact, explains a region's or territory's comparative advantage in terms of four major factors, each of which needs attention in order to move ahead of competing areas: enterprise strategy, structure and rivalry; demand conditions; spatial relationships with related and supporting industries; and resource and production factors (traditional or skill-based). Analyses in terms of a residential or local, face-to-face economy, which base territorial development on an increase in external revenue, have a different view of interregional disparities (Davezies, 2008).

Analyses of localised production systems (LPSs), which began in the 1970s, are also based on the observation of spatially differentiated development processes. Initiated by studies of Italian districts (Beccatini, 1990) and followed by studies of variations in different settings, ranging from the Milieus to the agrifood systems or LPSs to clusters, these analyses are based on the systemic nature of relationships maintained by actors who belong to and jointly shape a territory through their cooperation and common projects. It is here that we find the idea of development from below – so close to authors such as Stöhr (1986) – and a willingness to typify forms of development (the Italian districts; public-based systems; systems based around large companies; or based on innovation, etc.) (Markusen, 1996), but very little analysis on the development processes themselves or of their dynamics.

The notion of the New Economic Geography (NEG), conceived by Krugman (1991) and popularized by authors such as Fujita, Thisse and Ottaviano (Fujita and Thisse, 1997, 2001;

Ottaviano and Thisse, 2004), then formalized the significant probability of occurrence of phenomena of spatial polarization and concentration of activities. Questions then arise of the spillover effect of an activity at the regional level (e.g. spillover effect of construction), of the reciprocal impact of the locations of enterprises and those of their workers/consumers, and the ability to lower transportation costs, which only reinforces polarization processes to the detriment of peripheral areas.

8.2.3 Development as a dynamic process linked to innovation

The third and final research category is based on the idea that regional or territorial development is closely linked to the occurrence of dynamic ruptures with the past due to innovative or creative processes. This explains the varying speeds and amounts of development of different regions or territories (Dunford, 1993; Scott and Storper, 2003). Analyses of regional development based on processes of innovation and regulation, as well as some systemic approaches, thus conclude that local systems are subjected to successive phases of growth and stagnation, even of recession (Colletis *et al.*, 1999). These phases exacerbate or reduce inequalities between social classes, with the benefits of growth often being appropriated by certain groups or offshore businesses belonging to external capital. Above all, it is the internal shocks which can transform systems and lead to the appearance of spatial concentration of people and wealth, as well as of zones of social and spatial exclusion. Innovation, its creation and its dissemination are therefore at the heart of these approaches (Cooke and Morgan, 1998).

During the last decade, the analysis of spatial dynamics has been enriched by work rooted in evolutionary theory (Frenken and Boschma, 2007). It considers the uneven distribution of activities in space as resulting from largely contingent historical processes. The Evolutionary Economic Geography accords a predominant place to the entrepreneurial dimension, whether based on genealogy or on processes of emergence, growth, decline and cessation of business activity (Boschma and Frenken, 2011). The focus is mainly on the roles played by spin-offs and labour mobility in territorial development processes (Maskell, 2001) and on mechanisms for replicating routines within the local industrial system. Taking advantage of geographic, industrial and technological proximity between sectors (Torre, 2008) and of institutional mechanisms and network structures, these technologies spread by the snowball effect between the companies and technologically related industries, and eventually lock local systems into spatial dependencies on the growth path. This process works particularly well when the industries are emerging or are based on related technologies, with low cognitive distances being particularly conducive to the circulation of knowledge spillovers (Nooteboom, 2000).

8.3 Policies of development by innovation

One of the features of current development policies is their acceptance of local dynamism in innovation, production and knowledge transfer as one of the key factors in regional development. Hence, considerable efforts are made by regions and local communities in this domain. Policies to encourage innovation – a source of growing income – are today part of the toolbox of all policy makers, who see in them the ultimate argument for growth and development (Hall, 1994). These policies are based on the fact that gains from innovation are difficult to appropriate and thus require State intervention to meet any possible shortfalls in R&D spending. Such strategies have not only resulted in policies to promote high-tech activities (Goldstein, 2009) and major industrial programmes such as Airbus but are also considered relevant for rural or remote areas and SMEs which lack resources.

8.3.1 Technological innovation within poles of development

Approaches dealing with the role of innovation in the dynamics of territorial or regional development are based on taking into account the importance of R&D or innovation in local development. Partly inspired by Schumpeter's work, these approaches rely on the idea that innovations are key to development processes and that R&D efforts and incentives for innovation can play an important role in the establishment and success of the dynamics of growth. It is often a matter of a systemic approach, one which emphasizes the role played by innovation transfer and dissemination at the local level (Autant-Bernard *et al.*, 2007; Feldman, 1994). It also underlines the importance of face-to-face relations and of expansion phases by setting up of spin-offs and via support of creative efforts (nurseries, incubators, etc.). The engine of development is thus found in the presence of localized spillovers of innovation or knowledge, which spread within the local system and can give rise to very competitive local systems such as technology hubs or competitive clusters. It is innovation that powers development and differentiates dynamic systems from those that are not.

Advocating the concentration of industrial investment in clearly identified clusters is now a dominant feature of European policies but one limitation is due to these policies' linear design, which ignores the importance of feedback loops and uncertainty in innovation processes. Such approaches lead to rather poor results insofar that they omit the geographic concentration of R&D and innovation in a few regions and are unaware of the use of new knowledge outside the areas being covered. Moreover, pick-the-winner policies aimed at selecting areas most conducive to innovations and the sectors most likely to create new-economy jobs (biotechnology, nano-technology) can see their usefulness and relevance being called into question (Boschma, 2009). Besides the fact that it is impossible to predict future growth regions or successful sectors since new industries are often the results of spontaneous processes rather than of planned interventions, these policies lead to the adoption of the same activities everywhere whereas industrial and innovation systems are very different and often incomplete (Camagni, 1995). The phenomena of inertia and lock-in thus lead the great majority of regions to fail to develop these industries, resulting in huge losses of public resources.

These analyses draw support from the changed perception of innovation processes: from a purely linear model to the interactive one (Lundvall, 1992). Whereas the linear model, based on the Taylorist structure of production, described innovation as an unwavering process going from an initial idea to production to commercialization, the interactive model emphasizes the interactive and iterative nature of innovation between closely linked organizations at various stages of its development. Innovation is thus considered a social endeavour taking shape in a diversity of geographic configurations (Wolfe and Gertler, 2002). The linear model describes a spatial division of work based on a specialized functional hierarchy, with some regions benefitting from the positive effects in terms of income and growth due to their positioning and specialization in R&D activities. In contrast, the interactive model accords greater importance to the close relationships between knowledge users and knowledge creators through their geographical proximity and/or ICTs. Consequently, territorial institutional contexts are keys to explaining the potential and success of innovations (Bonaudo *et al.*, 2010; Puttilli and Tecco, 2010), with some areas proving to be much better than others in producing or adapting innovations (Malecki, 1997) (Box 8.1).

8.3.2 Innovation through knowledge creation

More recent work highlights the central role played by knowledge and its implications for territorial and regional development in association with innovation processes. According to these studies, development can be understood as the transformation of a set of assets consisting of products poorly developed and exploited by an under-qualified workforce into a set of knowledge-based assets exploited by skilled labour, with information regarded as an essential raw material (Lundvall and Maskell, 2000). Learning ability is thus revealed to be essential to the adaptive potential of territories and regions for their development. Learning is considered a collective, social and geographical process which brings about an improvement in individual or organizational understanding and capacities.

Some studies put emphasis more on the tension between individual representation and decision making and collective innovation, thus bringing the processes for creating and disseminating knowledge to the fore in the analysis. In this perspective, approaches based on territorially rooted communities of practices are marked by the use of an original conceptual framework to highlight the importance of routines and networks. Such approaches are similar to work on creative cities (Cohendet and Simon, 2008) and on evolutionary economic geography (Frenken and Boschma, 2007).

Finally, interdependent non-market relationships between institutions are key to a territory's or region's performance as measured by innovation, productivity growth and development. Relationships of trust – as well as high levels of tacit knowledge and the existence of routines

Box 8.1. The localized agrifood systems.

The approach of localized agrifood systems (LAS), originated at the end of the last century following the observation of organized exchanges and relationships between local actors involved in agricultural production or agrifood activity. This concept includes 'all production and service structures (farm, agrifood industry and businesses, catering businesses, etc.) linked by their characteristics and functioning to a specific territory. The environment, products, people, their institutions, their expertise, their food habits and their relationship networks all combine in a territory to form an agrifood structure at a given spatial scale (Muchnik, 1996).

This concept, which quickly became a major success with a section of the scientific community and with policy makers and public authorities (Muchnik and De Sainte Marie, 2010), is obviously an extension of the concept of localized production and innovation systems and other clusters to the agrifood production domain. The LASs are often found in rural areas, especially in developing countries, where the organization of local actors in production processes is based on local ties and sharing of skills and techniques (Sanz Cañada, 2010).

In addition, we note the significant references to technical aspects of production of goods, which is closely associated with the social context. We cannot take into account the production methods and related techniques without considering the modalities of the actors' social structuring, as well as the joint construction of social links and of the technical determinants in the action. We note, however, that these systems still remain orphans as far as a truly analytical assessment is concerned: they are interesting but lack substance. In fact, it is not possible to encapsulate them in a dominant or determinate theoretical approach even if the notions of economy of proximity or common goods seem to promise interesting developments in the coming years (Perrier-Cornet, 2009).

The question of the scale at which the innovation process takes place in association with the dynamics of development is also an essential element of the debate. Based on work on national innovation systems (Lundvall, 1992; Nelson, 1993; Freeman, 1995; Amable *et al.*, 1997), studies have been conducted on how these systems are deployed at the regional scale. They have sought to understand under what conditions local and regional networks and institutional mechanisms were more or less favourable to innovation and what were the conditions propitious to their adaptation and permanence over time (Lundvall and Maskell, 2000). These studies resulted in approaches of regional innovation systems (Cooke and Morgan, 1998) seeking to find ways to anchor innovations in territories and attempting to identify conditions leading to efficient and successful systems. This research insists on the importance of the presence of certain elements such as physical and technological infrastructure, R&D links between industry and universities, highly qualified workforce available on the local labour market and the existence of venture capital mechanisms. Also necessary are less tangible factors relating to the local social environment such as local know-how, a regional technical culture and proximity to collective cognitive frameworks. The role of regional and local institutional mechanisms appears therefore essential to reduce uncertainty and to support coordination and collective action conducive to innovation processes. Efficient systems are thus characterized by a high level of local interactions and interdependent relationships where innovation is supported and encouraged by public or private organizations.

– determine the structure of local mechanisms of cooperation and coordination. They can then be viewed as relational resources conducive to an increase in learning abilities and to the creation of benefits that other territories will find hard to replicate. In such a perspective, urban spaces and, more generally, urban territories are considered favourable to innovation and to knowledge creation due to the cognitive externalities they can generate (Scott and Storper, 2003).

The recognition of the role of innovation, knowledge and learning in the processes of regional and territorial development has had an impact on the evolution of development policies, which are now most often characterized by a set of infrastructure-oriented interventions (transport, high-speed telecommunications, etc.). These policies also extend support to less tangible elements such as network structuring and knowledge transfers in order to strengthen collective capacities of knowledge creation and learning. The challenge then remains to build assets that are endogenous to the territory. This is an objective that requires the mobilization of local forces in an interactive framework where the logic of experimentation (marked by an acceptance of the trial-and-error method) takes precedence over the implementation of predefined solutions, notwithstanding the constraints of public finances. This is why such mechanisms of public intervention are best assessed in the context of their construction rather than being assigned a universal value. Nevertheless, any examination of strategies pursued at the territorial or regional level (in addition to within a same national framework) shows the relatively low creativity of solutions put in place and the difficulty of most territories to differentiate themselves clearly and sustainably.

8.3.3 Towards territorial innovation?

Even though there has been undeniable progress over the last twenty years in the understanding of links between innovation, knowledge, learning and regional development, the theoretical models are still characterized by the diversity and weakness of their conceptualization and formalization, as well as by an unfortunate lack of clarity in messages destined for decision makers seeking to improve public policies. Often based solely on high-tech activities, oriented by technology and by a market-focused corporate culture, these proposals narrow the field of innovation to the most technological of dimensions. In this way, they neglect not only incremental innovations but also ignore many territories which do not adhere to high-tech principles but are still characterized by other sorts of vibrant innovation activities (social, organizational, institutional, etc.). Furthermore, apart from a facade of semantic unity based on their underlying concepts, these analytical models represent, in reality, different visions of the dynamics of innovation – hence the difficulty in establishing a clear theoretical framework.

A way forward on these issues, and in particular on including the question of innovation in an analysis that encompasses all territories, including rural ones, would be to broaden the debate to take into account the concept of territorial innovation in all its dimensions.

Such a debate should lead to an improved understanding of the progress of humanity at the territorial scale (Moulaert and Sekia, 2003) and to permit analysis of innovation models actually useful to local communities. Some approaches, for example the work of the Group for European Research on Innovative Environments (GREMI) on the concept of the innovative milieus (Camagni and Maillat, 2006), have investigated the concept of territorial innovation in the most rural or underdeveloped territories based on organizational innovations and on the mobilization of local populations. The rules for collective action and institutional mechanisms are then considered as factors explaining innovative territorial dynamics. Innovation is viewed as a social construct conditioned by the geographical context in which it occurs; rooted in practices, it is therefore necessarily located in the space. The issue of territorial innovation is also addressed by the emerging fields of social and solidarity-based economy and sustainable development (Zaoual, 2008). New concepts have been created such as that of social innovation (Hillier *et al.*, 2004; Klein and Harrison, 2007) which describes a set of corporate innovative practices in response to social needs which have been little met or unmet and/or implementing processes to incorporate an approach for social transformation over time. These initiatives show the prominent role played by territories as crucibles of new forms of organization and of innovative partnerships, both in urban and rural areas (Box 8.2).

8.4 What form of governance to help innovation emerge in rural and periurban areas?

Originally mainly centred around economic aspects, the analysis of the development process has gradually opened itself to the question of innovation by considering the interplay of local social and institutional relationships as well as the interactions and overlaps between geographical scales and levels. This increased complexity requires the issue of territorial governance to be addressed not only with an objective of helping innovative processes to emerge but also of incorporating the various aspirations and wishes of the local populations and to link them with overall policies and regulations.

Territorial governance processes are today undergoing intense upheavals. These latter shape the phases of territorial innovation and thus constitute an engine of development and growth in rural or urban territories. Such governance mechanisms can be viewed as laboratories of change because they accompany and sometimes anticipate the changes underway in the territories by giving them shape, by helping maintain a dialogue and expressions of opposition and by preventing violent confrontations or failures of development due to sluggishness or expatriation. These changes are embodied in the opposing and twin forms of conflict and consultation which constitute the modes of expression and the vehicles of transmission of ongoing innovations at the territorial level.

Box 8.2. An example of socio-territorial innovation: the agroecological transition in Brazil.
Marc Piraux, Philippe Bonnal, Luciano Silveira, Paulo Diniz and Ghislaine Duque

The agroecological transition of agricultural production systems is a complex process involving technical, social and institutional changes. The Brazilian Semi-Arid, a region where this transition was initiated over two decades ago, is an interesting example of innovation. Based on a desire to solve social problems caused by water shortages, further exacerbated by the inadequacy of technical solutions proposed by the government and the interventionist methods of a relatively centralized State, this transition led to a series of changes characteristic of social and territorial innovation: (1) they fulfilled the needs of local populations, an important condition of social innovation, (2) they strengthened the learning processes to develop capacities for a shared understanding of agro-ecosystems and to valorise socio-productive practices with little or no social status, with these practices constituting the basis of discussions on setting up an alternative project, (3) they were based on the social movement of experiments conducted by the farmers themselves, which led to a consolidation of social networks and, finally, (4) they valorised a multi-scale process of governance of agricultural innovations that promotes social and political integration of farmers. Even though rooted in the territory, these innovations also led to the creation and setting up of mechanisms (such as the Semi-Arid Articulation) and of public policies specific to the region.

These innovations have called into question the ability to improve the public sphere of development activities. Institutional experimentation was therefore essential. Institutional innovation shown by the State is in the form of having been able to transform the territorialized experiences of social actors into public policies. In this sense, it can be seen as an appropriation by the State of an innovation developed at the local scale. The limited success of this transformation is explained by the peripheral and piecemeal character of institutional change made within the State apparatus. The trajectories of social and institutional innovations are part of the very power relationships they seek to change. It appears that the sustainability of local socio-territorial innovations depends also on a complementary movement of harmonization of standards within the State itself (Piraux *et al.*, 2010; Bonnal and Piraux, 2010).

8.4.1 Consultation and negotiation to define a shared vision

To begin with, this concerns negotiation mechanisms, in particular those of consultation and their implementations at the local level. According to Beuret (2006), we can distinguish between different types of operations, characterized by increasing levels of involvement, that can be called upon within participatory approaches and which contribute at various levels to the territorial governance processes. Communication methods are used to convey messages and to obtain public support for proposals. Instead of relying on the balance of power, these methods can be used as part of participatory approaches, for example when it is a matter of convincing some groups that it is in their interest to participate. Information can be used to transmit data that would allow target individuals or groups to form an opinion and to participate in discussions. The actors' views can be ascertained via consultations but without any express guarantee that they will be accepted. Dialogue can draw participants closer together and lead to the establishment of a common language and references. Consultations encourage joint action and decision making and can be used to build a collective vision or goal and to set up joint projects. Finally, negotiations can be used to reach a decision acceptable to all participants.

In recent decades, these mechanisms have resulted in inventions and interventions of various kinds, all with the common purpose of facilitating the implementation of the consultation paradigm. The work of Ostrom (1990, 2005) is a successful example in creating mechanisms for governance of shared natural resources through the prism of property rights as defined by local communities. Nevertheless, it must be admitted that, on the whole, these mechanisms do not seem to be fully stabilized; in fact, they have set off debates and generated many controversies on their utility (Blatrix *et al.*, 2007; Mermet and Berlan-Darqué, 2009). A relative consensus has, however, emerged to acknowledge that various forms of participation by private or semi-public actors in debates or in public decision making does lead to more harmonious and democratic territorial governance processes. The result is a number of territorial governance mechanisms and tools, as the one presented in Box 8.3. Examples from France are the 1983 Bouchardeau Act and the 2002 Law on local democracy; increasing complexity of the decision making process relating to public infrastructure projects with the declaration of public utility, public hearings, and the setting up of the National Public Debate Commission; consultations before the creation or revision of urban master plans; and consultative commissions on local public services and utilities.

The consultation processes, characterized by a cooperative intent, form an important laboratory of coordination for improved territorial governance. The collective construction of these processes, based on the establishment of a structured and sustainable relationship between actors willing to share information, discuss problems or specific issues in order to agree on common objectives and possible collective action (Bourque, 2008), distinguishes them from other forms of cooperation and public-action participation. This consultative approach therefore encompasses 'processes of collective construction of visions, goals and joint projects in order to act or decide together' (Beuret, 2006). It can also be used by a third-party actor, such as an agent of development, to encourage coordination between various parties. It takes shape on stages – or arenas – around which revolve exchanges between groups of persons and entities characterized by the same actions relating to the subject under discussion and by the same attitudes and stances. In its history, the consultation process has often been subject to one or more controversies but the fact remains that its script is not written in advance and has to be developed on the fly as it follows a path of consultation.

Box 8.3. Co-construction of an analytical model and a guide for setting up territorial governance.
Hèlène Rey-Valette and Eduardo Chia

Within the framework of the Gouv.Innov[1] project (Governance of territories and rural development: an analysis of organizational innovations), we analyzed the dynamics of territorial and periurban governance in sustainable and integrated management situations and mechanisms. Particular attention was paid to the role of agricultural actors. It was a matter of observing, analyzing and monitoring the setting up and functioning of governance mechanisms. One operational objective was to offer a guide to local development actors for implementing territorial governance. This guide was jointly written by the multidisciplinary research team of the Gouv.Innov project and a panel consisting of development officers and farmer representatives. The starting point of the guide was a generic analysis of territorial governance. It permitted the taking into account of the diversity of mechanisms as well as of their characters - historically based on territorial processes - and of their spatial embedding and multi-scale dynamics. Four stages can be distinguished, corresponding to complementary analysis categories. They can be undertaken sequentially or, for limited requirements, in isolation. They are:

- improving understanding (stage 1, knowledge of actors, institutions and procedures);
- analyzing (detailed examination of functional interactions (stage 2) and territorial interactions (stage 3), using surveys, interviews or forms of participant observations);
- evaluating products and effects of processes (stage 4), corresponding to the deepest level that incorporates a logic of reflexivity with the greatest information requirements.

The guide (http://www.lameta.univ-montp1.fr/ggov) that resulted from this collective work stresses the 'operational' aspect of the definition of governance by proposing ways to re-engineer territorial governance at several levels. It does so by suggesting measures to strengthen (1) the preparatory phases, (2) the support extended to the actors, (3) the guidance and coordination of public action, (4) the evaluation of governance mechanisms and (5) the institutionalization for the sake of durability. It also includes several on-the-ground examples and is a true 'tool' for reflexivity and support for the engineering of territorial governance (Rey-Valette *et al.*, 2011).

8.4.2 The role of conflicts in the processes of innovation

Our research into conflicts in rural and periurban areas (Torre, 2010; Torre *et al.*, 2006) shows that this dimension is also key in processes of territorial management, regional development or the governance of various local activities. It appears in the form of litigation, media events or violent protests. In most cases, land-use conflicts are not blind oppositions or purely

[1] Project funded by the Languedoc-Roussillon 'For and On Regional Development' (PSDR) research programme.

egoistical in origin but constitute a way of initiating discussions on the issues and paths of territorial development and of influencing decisions by participating in processes underway from which one had been excluded (Dowding *et al.*, 2000). That is why they have a bearing, either on the decisions on land use and management (arbitrated negotiation) or on the composition and representativeness of the bodies responsible for taking decisions (arbitral negotiation). The conflict thus becomes an integral part of the deliberative process at the local level by allowing an expression of local democracy and the re-inclusion of participants who were forgotten or deliberately excluded during earlier project development stages.

Land-use conflicts thus constitute one form of resistance and expression of opposition to decisions that leave part of the local population unsatisfied (Darly and Torre, 2013). Some local innovations, whether technical or organizational in nature, give rise to resistance which can turn into conflict. Major changes requiring reconfiguration of the use of space (creation of transport, energy or waste-processing infrastructure, new urban master plans, territorial or environmental zoning, etc.) generate conflicts whose spatial and social extent can quickly grow. Conflicts are signals of social, technological and economic changes, indicators of novelty and innovations. They demonstrate the opposition aroused by the latter, lead to discussions on their implementations and their possible (non-)acceptability as well as on the adoption of governance procedures and their transformation under the influence of the dynamics of change. All changes encounter opposition or resistance of varying relevance and justification. But it would however be simplistic to see this resistance as a systemic sign of reactionary opposition to change because, in a number of cases, they are more a reflection of differences over the direction taken by the new initiatives that are being imposed on the public than of a stubborn desire to maintain the status quo. During these phases of conflict, social and interest groups tend to reconstitute themselves and may even undergo technical or legal changes. Once a conflict ends, it leaves behind new local agreements, new modes of governance, new configurations of discussion forums as well as new technical procedures (changes in direction, various adjustments, changes in urban planning documents, etc.), all arrived at during the negotiations. Harbingers of territorial innovation, conflicts are thus both the result as well as the cause of territorial changes.

Territorial governance is therefore not limited to an idyllic vision of economic and social relationships, i.e. to forms of cooperation and common constructions. It also involves interactions between forces favouring cooperation and those pushing towards conflict (Torre and Traversac, 2011). Far from resembling a smoothly flowing course, territorial development processes and their implementations over time are made up not only of processes of negotiation, collaboration or appeasement but also of more lively or confrontational phases during which some groups or category of actors face off, sometimes violently, in order to define the way forward and to make choices. The process of territorial governance therefore presents two complementary aspects whose mutual importance varies with time and situations. It feeds on these opposing tendencies (Glazer and Konrad, 2005), with their synthesis and combination revealing paths of territorial development.

8.5 Conclusions

Today, many authors consider that a new paradigm of rural development is being created which is independent of the agro-industrial and hygienic model of production based on the use of chemical inputs and sanitary control of products. It builds a representation of rural spaces that differs from one of dependence on urbanization (Röling and de Jong, 1998; Marsden, 2006). Additionally significant is the rise of environmental and sustainable-development issues, which are impacting strongly the design of rural activities, especially agricultural activity, as well as influencing public policies through their local implementations, in particular via zoning processes (Natura 2000, habitat directives, green and blue belts, etc.).

This new paradigm emerges both in the local actors' practices and procedures and in public policies, with rural development being seen as a multi-level, multi-actor and multi-faceted process (Van der Ploeg *et al.*, 2000). Multi-level in the diversity of policies and institutions designed to address the issues of rural development, as well as the evolution of the agriculture-society relationship, taking into account the production of public goods, the construction of a new agricultural production model incorporating interactions between agriculture and other activities and the combining of activities at the enterprise scale in rural areas. Multi-actor because of the interactions between farmers and other rural-area actors and because of the rural development policies designed to bring about new links between the local and the global. They can however also be used to restore the legitimacy of local elites or to play on clientilist interests. Finally, multi-faceted because rural development unfolds into a range of differentiated practices, some of which are emerging and sometimes interconnected (landscape management, nature conservation, agritourism, organic farming, specific agricultural products, short supply chains, etc.) so that elements considered redundant in modernist paradigms acquire new roles in farm-to-farm relationships and in those between farmers and the urban population.

References

Amable, B., Barré, R. and Boyer, R. (eds.), 1997. Les systèmes d'innovation à l'ère de la globalisation. Economica, Paris, France.

Autant-Bernard, C., Mairesse, J. and Massard, N., 2007. Spatial knowledge diffusion through collaborative networks. Papers in Regional Science, 86(3), 341-350.

Becattini, G., 1990. The Marshallian industrial districts as a socio-economic notion. In: Pyke, F., Becattini, G. and Sengenberger, W. (eds.) Industrial districts and inter-firm cooperation in Italy. International Institute of Labour Studies, Geneva, Switzerland, pp. 37-51.

Beuret, J.E., 2006. La conduite de la concertation. Pour la gestion de l'environnement et le partage des ressources. L'Harmattan, Paris, France.

Blatrix, C., Blondiaux, L., Fourniau, J.M., Heriad-Dubreil, B., Lefebvre, R. and Revel, M., 2007. Le débat public: une expérience française de démocratie participative. La Découverte, Paris, France.

Bonaudo, T., Coutinho, C., Poccard-Chapuis, R., Lescoat, P., Lossouarn, J. and Tourrand, J.F., 2010. Poultry industry and the sustainable development of territories: what links? what conditions? In: Innovation and Sustainable Development in Agriculture and Food – ISDA 2010, Montpellier, France. Available at: http://hal.archives-ouvertes.fr/hal-00522800/fr/.

Bonnal, P. and Piraux, M., 2010. Actions publiques territoriales en milieu rural et innovations l´exemple du territoire de la Borborema et de l'articulation du semi-aride au Brésil. In: Innovation and Sustainable Development in Agriculture and Food – ISDA 2010, Montpellier, France. Available at: http://hal.archives-ouvertes.fr/hal-00522109/fr/.

Boschma, R. and Frenken, K., 2011. The emerging empirics of evolutionary economic geography. Journal of Economic Geography, 11, 295-307.

Boschma, R., 2009. Evolutionary economic geography and its implications for regional innovation policy. Papers in Evolutionary Economic Geography, 09.12.

Bourque, D., 2008. Concertation et partenariat. Entre levier et piège du développement des communautés. Collection Initiatives, Presses de l'Université du Québec, Quebec, Canada.

Camagni, R. and Maillat, D. (eds.), 2006. Milieux innovateurs. Théorie et politiques. Anthropos, Economica, Paris, France.

Camagni, R., 1995. The concept of innovative milieu and its relevance for public policies in European lagging regions. Papers in Regional Science, 74(4), 317-340.

Cohendet, P. and Simon, L., 2008. Knowledge intensive firms, communities and creative cities. Oxford University Press, Oxford, UK.

Colletis, G., Gilly, J.-P., Leroux, I., Pecqueur, B., Perrat, J., Rychen, F. and Zimmermann, J.B., 1999. Construction territoriale et dynamiques productives. Sciences de la Société, 48, 35-54.

Cooke, P. and Morgan, K., 1998. The associational economy: firms, regions and innovation. Oxford University Press, Oxford, UK.

Darly, S. and Torre, A., 2013. Land-use conflicts and the sharing of resources between urban and agricultural activities in the Greater Paris Region. Results based on information provided by the daily regional press. In: De Noronha Vaz, T., Van Leeuwen, E. and Nijkamp, P. (eds.), Towns in a rural world. Routledge, London, UK. Available at: http://www-sre.wu.ac.at/ersa/ersaconfs/ersa10/ERSA2010finalpaper1325.pdf.

Davezies, L., 2008. La République et ses Territoires. Editions du Seuil, Paris, France.

Dowding, K., John, P., Mergoupis, T. and Van Vugt, M., 2000. Exit, voice and loyalty: analytic and empirical developments. European Journal of Political Research, 37, 469-495.

Dunford, M., 1993. Regional disparities in European Community: evidence from the REGIO databank. Regional Studies, 27(8), 727-743.

Feldman, M., 2000. Location and innovation: the new economic geography of innovation, spillovers and agglomeration. In: G.L. Clark, M. Feldman and G. Gertler (eds.), The Oxford Handbook of Economic Geography. Oxford University Press, Oxford, UK, pp. 373-394.

Freeman, C., 1995. The 'national system of innovation' in historical perspective. Cambridge Journal of Economics, 19, 5-24.

Frenken, K. and Boschma, R., 2007. A theoretical framework for evolutionary economic geography: industrial dynamics and urban growth as a branching process. Journal of Economic Geography, 7(5), 635-649.

Fujita, M. and Thisse, J.F., 1997. Economie géographique, problèmes anciens et nouvelles perspectives. Annales d'Economie et de Statistique, 45, 37-87.

Fujita, M. and Thisse, J.F., 2001. Economie et marché. Cahiers d'Economie et de Sociologie Rurale, 58-59, 11-57.

Glazer, A. and Konrad, K.A. (eds.), 2005. Conflict and governance. Springer Verlag, Berlin, Germany.

Goldstein, H., 2009. Theory and practice of technology-based economic development. In: Rowe, J.E. (ed.), Theories of local economic development. Ashgate, Burlington, UK, pp. 237-264.

Hall, P., 1994. Innovation, economics and evolution. Harvester Wheatsheaf, New York, NY, USA.

Hillier, J., Moulaert, F. and Nussbaumer, J., 2004. Trois essais sur le rôle de l'innovation sociale dans le développement territorial. Géographie, économie, société 6, 129-152.

Johansson, B., Karlsson, C. and Stough, R.R., 2001. Theories of endogenous regional growth. Springer, Heidelberg, Germany.

Klein, J.L. and Harrison, D. (eds.), 2007. L'innovation sociale. Emergence et effets sur la transformation des sociétés. Presses de l'Université du Québec, Quebec, Canada.

Krugman, P., 1991. Geography and trade. MIT Press, Cambridge, MA, USA.

Lundvall, B-A. (ed.), 1992. National innovation system: towards a theory of innovation and interactive learning. Pinter, London, UK.

Lundvall, B-A. and Maskell, P., 2000. Nation states and economic development: from national systems of production to national systems of knowledge creation and learning. In: Clark, G.L., Feldman, M. and Gertler, G. (eds.), The Oxford Handbook of Economic Geography. Oxford University Press, Oxford, UK, pp. 353-372.

Malecki, E., 1997. Technology and economic development: the dynamics of local, regional and national competitiveness, 2nd ed. Addison Wesley Longmann, London, UK.

Markusen, A., 1996. Sticky places in slippery space: a typology of industrial districts. Economic Geography, 72 (2), 294-314.

Marsden, T., 2006. Pathways in the sociology of rural knowledge. In: Cloke, P., Marsden, T. and Mooney, P. (eds.) The Handbook of Rural Studies. Sage Publications, London, UK, pp. 3-17.

Maskell, P., 2001. The firm in economic geography. Economic Geography, 77(4), 329-344.

Mermet, L. and Berlan-Darqué, M. (eds.), 2009. Environnement: décider autrement. Nouvelles pratiques et nouveaux enjeux de la concertation. L'Harmattan, Pais, France.

Moulaert, F. and Sekia, F., 2003. Territorial innovation models: a critical survey. Regional Studies, 37(3), 289-302.

Muchnik, J. and De Sainte Marie, C. (eds.), 2010. Le temps des syal. techniques, vivres et territoires. Coll. Update Sciences & Technologies, Ed. Quae, Paris, France.

Muchnik, J., 1996. Systèmes agroalimentaires localisés: organisation, innovations et développement local. Proposition issue de la consultation du Cirad Stratégies de recherche dans le domaine de la socioéconomie de l'alimentation et des industries agroalimentaires, Cirad, Montpellier, France.

Nelson, R. (ed.), 1993. National innovations systems. A comparative analysis. Oxford University Press, Oxford, UK.

Nooteboom, B., 2000. Learning and innovation in organizations and economies. Oxford University Press, Oxford, UK.

Ostrom, E., 1990. Governing the commons: the evolution of institutions for collective action. Cambridge University Press, Cambridge, UK.

Ostrom, E., 2005. Understanding institutional diversity. Princeton University Press, Princeton, NJ, USA.

Ottaviano, G. and Thisse, J.F., 2004. Agglomeration and economic geography. In: Henderson, J.V. and Thisse, J.F. (eds.), Handbook of Regional and Urban Economics, edition 1, vol. 4. Elsevier, Amsterdam, the Netherlands, pp. 2563-2608,

Perrier-Cornet, Ph., 2009. Les systèmes agroalimentaires localisés sont-ils ancrés localement? Un bilan de la littérature contemporaine sur les Syal. In: Aubert, F., Piveteau, V. and Schmitt, B. (eds.), Politiques agricoles et territoires. Quae, Versailles, France.

Perroux, F., 1961. L'économie du XXe siècle. Presses Universitaires de Grenoble, Grenoble, France.

Piraux, M., Silveira, L., Diniz, P. and Duque, G., 2010. La transition agroécologique comme une innovation socio-territoriale. In: Innovation and Sustainable Development in Agriculture and Food – ISDA 2010, Montpellier, France. Available at: http://hal.archives-ouvertes.fr/hal-00512788/fr/.

Porter, M.E., 1985. Competitive advantage. The Free Press, New York, NY, USA.

Porter, M.E., 1990. The competitive advantage of nations. The Free Press, New York, NY, USA.

Puttilli, M. and Tecco, N., 2010. Implications of biodiesel production chains in restructuring rural space. In: Innovation and Sustainable Development in Agriculture and Food – ISDA 2010, Montpellier, France. Available at: http://hal.archives-ouvertes.fr/hal-00520767/fr/.

Regional Science Policy and Practice, 2011. Special issue on innovation and creativity as the core of regional and local development policy. Regional Science Policy and Practice, 3 (3).

Rey-Valette, H., Chia, E., Soulard, C., Mathe, S., Michel, L., Nougaredes, B., Jarrige, F., Maurel, P., Clement, C. and Martinand P., Guiheneuf, P.Y. and Barbe, E., 2010. Innovations et gouvernance territoriale: une analyse par les dispositifs. In: Innovation and Sustainable Development in Agriculture and Food – ISDA 2010, Montpellier, France. Available at: http://hal.archives-ouvertes.fr/hal-00520264/fr/.

Röling, N. and De Jong, F., 1998. Learning: shifting paradigms in education and extension studies. Agricultural Education and Extension, 5(3), 143-161.

Sanz Cañada, J., 2010. Territorial externalities in local agro-food systems of typical food products the olive oil protected designations of origin in Spain. In: Innovation and Sustainable Development in Agriculture and Food – ISDA 2010, Montpellier, France. Available at: http://hal.archives-ouvertes.fr/hal-00530963/fr/.

Scott, A.J. and Storper, M., 2003. Regions, globalization, development. Regional Studies, 37(6-7), 579-593.

Solow, R.M., 2000. Growth theory: an exposition. Oxford University Press, New York, NY, USA.

Stöhr, W.B., 1986. Regional innovation complexes. Papers in Regional Science, 59, 29-44.

Torre, A. and Traversac, J.B., 2011. Territorial governance. Springer Verlag, Heidelberg, Germany.

Torre, A. and Wallet, F., 2013. The intriguing question of regional and territorial development in rural areas. Analytical variations and public policies. European Planning Studies, in press.

Torre, A., 2008. On the role played by temporary geographical proximity in knowledge transfer. Regional Studies, 42(6), 869-889.

Torre, A., 2010. Conflits environnementaux et territoires. In: Zuindeau, B. (ed.), Développement Durable et Territoire. Presses Universitaires du Septentrion, Villeneuve-d'Ascq.

Torre, A., Aznar, O., Bonin, M., Caron, A., Chia, E., Galman, M., Guérin, M., Jeanneaux, Ph., Kirat, Th., Lefranc, Ch., Melot, R., Paoli, J.C., Salazar, M.I. and Thinon, P., 2006. Conflits et tensions autour des usages de l'espace dans les territoires ruraux et périurbains. Le cas de six zones géographiques françaises. Revue d'Economie Régionale et Urbaine, 3, 415-453.

Van der Ploeg, J.D., Renting, H., Brunori, G., Knicken, K., Mannion, J., Marsden, T., De Roest, K., Sevilla Guzman, E. and Ventura, F., 2000. Rural development: from practices and policies towards theory. Sociologia Ruralis, 40, 391-408.

Wolfe, D.A. and Gertler, M., 2002. Innovation and social learning: an introduction. In: Gertler, M. and Wolfe, D.A. (eds.) Innovation and social learning: institutional adaptation in an era of technological change. Palgrave Macmillan, Basingstoke, UK, pp. 1-24.

Zaoual, H. (ed.), 2008. Développement durable des territoires: économie sociale, environnement et innovations. Coll. marchés et organisations, L'Harmattan, Paris, France.

Part III. What implications for policy making and research?

Chapter 9. Agrobiodiversity : towards inovating legal systems

Juliana Santilli

9.1 Introduction

The innovative concept of 'agrobiodiversity' has emerged in the past 10-15 years at the intersection of biodiversity and agriculture, in an interdisciplinary context that involves various areas of knowledge (agronomy, anthropology, ecology, botany, genetics, conservation biology, etc.). It reflects the dynamic and complex relations among human societies, cultivated plants and domestic animals and the ecosystems in which they interact. Agrobiodiversity is directly associated with food security, health, social equity, hunger alleviation, environmental sustainability and rural sustainable development. In the same way as wild biodiversity, agrobiodiversity has been considered in danger and in great need to be safeguarded through new legal instruments, at the international and national levels. After describing the rationales behind agrobiodiversity, this chapter highlights the various new legal tools applying to agrobiodiversity, which tend to adopt a more systemic view of the agroecosystem as whole, instead of focusing only on specific objects. These new legal instruments consider, for the first time, farmers´rights, agricultural systems as part of cultural heritage and the role of protected areas in promoting agrobiodiversity. The safeguarding of Globally Important Agricultural Heritage Systems is also discussed. There is a rich innovation process going on amongst lawmakers over the last two decades which aims to protect and valorise local and traditional agricultural systems and the rich agrobiodiversity that they encompass.

9.2 Agrobiodiversity: a concept under construction

Biodiversity or biological diversity – the diversity of life forms – covers three degrees of variability: diversity of species, genetic diversity (variability within the set of individuals of a given species) and ecological diversity, referring to different ecosystems and landscapes. The same is true of agrobiodiversity, which comprises the diversity of species (different species of cultivated plants, such as maize, rice, pumpkins, tomatoes, etc., called interspecific diversity), genetic diversity within a given species (different varieties of maize or beans, etc., called intraspecific diversity) and the diversity of cultivated ecosystems or agroecosystems (agroforestry systems, shift cultivation, home gardens, rice paddy fields, etc.). Local knowledge and culture are also integral parts of agricultural biodiversity, since it is agriculture, a human activity, that conserves biodiversity.

According to Cromwell *et al.* (2003), agricultural biodiversity includes (1) higher plants: crops, wild plants harvested and managed for food, trees on farms, pasture and rangeland species; (2) higher animals: domestic animals, wild animals hunted for food, wild and farmed fish; (3) arthropods: mostly insects including pollinators (e.g. bees, butterflies), pests (e.g. wasps, beetles), and insects involved in the soil cycle (notably termites); (4) other macro-organisms (e.g. earthworms); (5) micro-organisms (e.g. rhizobia, fungi, disease-producing pathogens). Agrobiodiversity's functions are related to sustainable production of food and other agricultural products, including providing the building blocks for the evolution or deliberate breeding of useful new crop varieties; biological support to production by means of, for example, soil biota, pollinators, and predators; and wider ecological services provided by agroecosystems, such as landscape protection, soil protection and health, water cycling and quality, and air quality (Cromwell *et al.*, 2003).

Agrobiodiversity is generally associated with crops (cultivated plants). However, Cromwell *et al.* (2003) point out that wild species are important nutritionally and culturally. Foods from wild species form an integral part of the daily diets of many rural households. Livestock diversity is also an important component of agrobiodiversity. Domesticated animals provide people not only with food but also with clothing, fertilizer and fuel (from manure), and draft power. Social and cultural forces are often the most important factors in diversifying livestock (and livestock production systems) and in developing distinctive breeds. Most local livestock breeds in rural environments are products of a community of breeders, and the effective management of animal genetic diversity is essential to global food security and sustainable development.

The Convention on Biological Diversity (CBD) does not contain a definition of agrobiodiversity, but, according to Decision V/5, adopted during the Fifth Conference of the Parties of CBD (COP-5), 'agricultural biodiversity is a broad term that includes all components of biological diversity of relevance to food and agriculture, and all components of biological diversity that constitute the agroecosystem: the variety and variability of animals, plants and microorganisms, at the genetic, species and ecosystem levels, which are necessary to sustain key functions of the agroecosystem, its structure and processes.' According to Decision V/5 of CBD, the special nature of agricultural biodiversity includes the following features:

a. Agricultural biodiversity is essential to satisfy basic human needs for food and livelihood security.
b. Agricultural biodiversity is managed by farmers. Many components of agricultural biodiversity depend on this human influence. Indigenous knowledge and culture are integral parts of the management of agricultural biodiversity.
c. There is a great interdependence among countries on the genetic resources for food and agriculture.
d. For crops and domestic animals, diversity within species is at least as important as diversity between species and has been greatly expanded through agriculture.
e. Because of the degree of human management of agricultural biodiversity, its conservation in production systems is inherently linked to sustainable use.

f. Nonetheless, much biological diversity is now conserved *ex situ* in genebanks or breeders' materials.
g. The interaction between the environment, genetic resources and management practices that occurs within agroecosystems contributes to maintaining dynamic agricultural biodiversity.

9.3 Agrobiodiversity and food security, nutrition, health, environmental sustainability and climate change adaptation

It is the diversity of cultivated plants and domestic animals, and their capacity to adapt to adverse environmental conditions (climate, soil, vegetation, etc.) and to specific human needs, that allows farmers to survive in the harshest regions of the world. It is the cultivation of various species and varieties that protects farmers, in many circumstances, from total loss of the harvest, in cases of pests, diseases, droughts, floods, etc. With monocultures, which have an extremely narrow genetic base, the opposite occurs: pests and diseases affect the only grown species and destroy the harvest.

Agrobiodiversity is also an essential part of making agriculture more sustainable. Sustainable agriculture is delivered when farming produces food and other agricultural products to satisfy human needs indefinitely without unsustainable impacts on the broader environment. This requires agriculture to avoid severe or irreversible damage to the ecosystem services (such as soil fertility, water quantity and quality, genetic variability, pollinators, etc.) upon which it depends and to have acceptable impacts on the broader environment (environmental stewardship). Sustainable agriculture will rely more and more on agricultural biodiversity, making the best use of the widest range of species and varieties to build greater resilience to external shocks and increase harvests, rather than relying upon external inputs (Bioversity International, 2010).

Agrobiodiversity is not only associated with sustainable food production, but also plays an essential role in the promotion of food and nutrition security and human health. A diversified diet – balanced in protein, vitamins, minerals and other nutrients – is recommended by nutritionists and is a fundamental condition for good health. An unbalanced diet may lead to what is called the 'hidden hunger', that is, the lack of essential micronutrients (vitamins and minerals) in the everyday diet. The effects of 'hidden hunger' are wide-ranging and may include delayed mental development, weakening of the immune system and loss of strength and energy. It is estimated that some 2 billion people are affected by a lack of iron, and around 800 million are deficient in vitamin A, the lack of which can in severe cases lead to blindness (GTZ, 2006). The best insurance against nutrient deficiencies is eating a varied diet, thereby ensuring an adequate intake of all the macro and micronutrients needed for good health.

Only diverse agricultural systems can offer more nutritive and balanced diets, and the fight against hunger and poverty must necessarily include more sustainable agricultural practices.

There is a direct relation between reduction in agricultural diversity and impoverishment of diets. Genetic erosion in the fields affects not only farmers but consumers as well. Agricultural production models have direct implications for human nutrition and health. 'Modern' agriculture and cultivation of few agricultural species favour standardization of dietary habits and loss of cultural appreciation of native species. Therefore, agrobiodiversity is directly associated to at least two of the Millennium Development Goals: the eradication of poverty and hunger and the achievement of environmental sustainability. In addition, it is essential for food security, which is achieved 'when all people, at all times, have physical and economic access to sufficient, safe and nutritious food to meet their dietary needs and food preferences for an active and healthy life' (FAO, 1996).

Finally, it is important to stress that agrobiodiversity is an essential component of international and national strategies to adapt or mitigate the impacts of global climate change. There is a close relationship between climate change, food security, and agrobiodiversity: it is diversity that makes it possible for species, varieties, and agroecosystems to adapt to changes and variations in environmental conditions. They can cope with future challenges, including those brought about by climate change, only if they have wide genetic variability. Interactions between agrobiodiversity and climate change are many, not least of which is that climate change is reducing the number of species and agricultural ecosystems. On the other hand, agrobiodiversity is also essential to face the impacts caused by global warming.

9.4 Agrobiodiversity and legal instruments

9.4.1 Agrobiodiversity and seed laws

Seeds – we shall use this term in a broad sense, to include all plant propagating materials – contain the entire life of a plant and constitute the basis of agrobiodiversity. Therefore, the impact of legal systems on agricultural diversity cannot be understood without an analysis of regulations on production, marketing, and utilization of seeds – the so-called 'seed laws'.

Seed laws aim to ensure the identity and quality of plant propagating materials, regulating how they are produced, used and marketed. They must not be confused with intellectual property (IP) laws or plant breeders' rights, which grant ownership rights over new plant varieties. However, seed laws and IP rights over plant varieties follow similar logics and share some common concepts.

Seed laws were adopted as a 'legal support' to the so-called process of 'modernization' or 'industrialization' of agriculture, whose main paradigms were productivism, standardization of agricultural products and fragmentation of the many stages involved in agricultural production. This new industrial paradigm promoted high-yield, homogeneous and stable plant varieties, highly dependent on external inputs. Analyzing the historical development of formal/commercial seed systems, Niels Louwaars (2007) explains that they emerged in industrialized countries in the second half of the nineteenth century. In the 1960s and 1970s, many developing countries also started to promote formal/commercial seed systems with wide support from international organizations. However, formal/commercial seed systems were never able to replace (totally) farmers' seed systems and their locally adapted seeds.

Bonneuil *et al.* (2006) explain that 'modernization' of agriculture promoted the idea that both plant breeding and seed production should be activities performed by specific professional sectors (plant breeders, agronomists, etc.). Farmers were seen as mere producers and consumers of seeds and other industrially produced agricultural inputs. This concept denied the role of farmers as innovators and holders of knowledge and practices that are crucial for the sustainability of agroecosystems and for on-farm conservation of agrobiodiversity. Seeds and varieties developed and produced by farmers, adapted to local conditions, were more and more replaced by 'modern' uniform varieties, and agricultural knowledge was increasingly produced off-farm, away from farmers. Mainstream concepts – that the homogeneous and stable variety is 'perfect' and the most appropriate for any farming system and that only professional scientists are capable of innovating in agriculture – were the basis for seed laws passed in the post-Green Revolution period, as Bonneuil *et al.* (2006) explain. Therefore, seed laws tend to be biased toward the so-called 'formal' seed system, and do not take into consideration the role of 'local' seed systems (also called 'informal'), which are managed by farmers themselves, and involve the production, multiplication, distribution, exchange, improvement and conservation of seeds at the local level (Box 9.1).

Seed laws must take into consideration the different needs of agricultural systems. They must recognize the value of different seed systems for different situations, and the important role of farmers' systems in the conservation and sustainable use of plant genetic resources for food and agriculture.

Seed regulations must be reviewed and adjusted to the need to promote on-farm and *in-situ* agrobiodiversity conservation. They must support the development and maintenance of diverse farming systems (and, for that purpose, they must consider the need of diversified seed systems), increase the range of genetic diversity available to farmers and promote the expanded use of local and locally adapted crops, varieties and underutilized species.

9.4.2 Agrobiodiversity and farmers' rights

The realization of farmers' rights is essential to ensure the conservation and sustainable use of agrobiodiversity. Farmers' rights include collective rights to land and agrarian reform, water, natural resources, energy, appropriate technology and education, health care, political participation and freedom of association, among others. These rights are intrinsically linked, but here we will focus on farmers' rights as set out in the preamble and in article 9 of the International Treaty on Plant Genetic Resources for Food and Agriculture (ITPGRFA). We believe that this instrument provides an important opportunity to construct and implement

Box 9.1. When farmers in poor communities combine participatory plant breeding and in-situ and ex-situ management. An illustration from Honduras, Central America.
Henri Hocde, Juan Carlos Rosas and Rodolfo Araya

In the central highlands of Honduras, perched at an altitude of 1000 m, poor communities benefiting little or not at all from research services (which tend to be focused on geographical areas more conducive to 'modern' production conditions) formed local agricultural research committees (called CIAL). With the support of NGO professionals and a few national plant breeder scientists, they launched themselves into the breeding of food crops (beans and maize). After many patient years of tenacious perseverance, of ongoing learning, of fighting doubts and of motivating themselves, they finally tasted success. As a result of their efforts, new varieties are now emerging and being cultivated, for example, Palmichal, Cedron, Nueva Esperanza, Capulin and Olotillo mejorado. At the same time, the CIALs established community seed banks to store their communities' – and neighbouring communities' – local varieties and the improved varieties that they are progressively perfecting (over 70 local bean varieties, 40 of maize, 20 improved ones of vegetables and trees, etc.). Samples of the varieties stored in these mountain communities are also stored in the national germplasm bank of the research centre which supports them. This latter is itself connected at the international level to other germplasm banks (CIAT in the case of beans, for example). Such an interconnected arrangement facilitates the *in situ* and *ex situ* preservation of agrobiodiversity.

The sequence of their work is illuminating: In 2005, the CIALs started identifying, cataloguing and inventorying the range of varieties that farmers in the region were sowing and began storing them in their banks. From these varieties, they identified those that could serve as parents for cross-breeding with varieties from the national bank. They then multiplied the seeds of the created varieties and supplied the farmers in their communities. Subsequently, they went out to the major cities in their areas and participated in biodiversity fairs to expose the reality of agrobiodiversity to a wider rural and urban audience: its precise meaning, its usefulness and the methods of managing it. Gradually, complementing the work of the community seed banks and in collaboration with them, some farmers, designated 'seed guardians', undertook the conservation of individual varieties. For these communities, it is these operations, taken as a whole as well in their sequence, that contribute to the management (and enrichment) of agrobiodiversity (Hocdé *et al.*, 2010).

farmers' rights at the national level, as it is the first legally binding international agreement that explicitly recognizes farmers' rights. These are the main provisions on farmers' rights, as set out in the article 9 of the ITPGRFA:

9.1. The contracting parties recognize the enormous contribution that the local and Indigenous communities and farmers of all regions of the world, particularly those in the centers of origin and crop diversity, have made and will continue to make for the conservation and development of plant genetic resources which constitute the basis of food and agriculture production throughout the world.

9.2. The contracting parties agree that the responsibility for realizing farmers' rights, as they relate to plant genetic resources for food and agriculture, rests with national governments. In accordance with their needs and priorities, each contracting party should, as appropriate, and subject to its national legislation, take measures to protect and promote farmers' rights, including:
   a. protection of traditional knowledge relevant to plant genetic resources for food and agriculture;
   b. the right to equitably participate in sharing benefits arising from the utilization of plant genetic resources for food and agriculture; and
   c. the right to participate in making decisions, at the national level, on matters related to the conservation and sustainable use of plant genetic resources for food and agriculture.

9.3. Nothing in this Article shall be interpreted to limit any rights that farmers have to save, use, exchange and sell farm-saved seed/propagating material, subject to national law and as appropriate.

Although the Treaty acknowledges that countries must adopt measures to protect farmers' rights, each country may decide which measures to adopt, and policies and actions listed in the Treaty are merely illustrative, allowing countries to adopt others. The preamble of the ITPGRFA refers explicitly to the rights that farmers have to 'save, use, exchange and sell farm-saved seed and other propagating material.' Article 9.3, however, affirms that 'nothing in this article shall be interpreted to limit any rights that farmers have to save, use, exchange and sell farm-saved seed/propagating material, subject to national law and as appropriate.' Although the preamble affirmatively recognizes these rights, Article 9.3 is neutral and sets forth only that the decision rests with each country, and according to its national law. Article 9.3 reflects the lack of consensus among countries that defended a positive recognition of farmers' rights to save, use, exchange and sell farm-saved seeds and countries that were against such a positive recognition, which could lead to restrictions on plant breeders' rights.

Article 9.3 does not, however, make any restrictions on the options that can be adopted by countries regarding the implementation of farmers' rights at the national level, even if it includes limitations on intellectual property rights over plant varieties. This is probably one of the most divisive issues regarding the implementation of farmers' rights. Neither the proposal submitted by the EU nor the one presented by the United States during the elaboration of the International Treaty on Plant Genetic Resources for Food and Agriculture made any reference to farmers' rights to save, use, exchange and sell farm-saved seeds.

For agrobiodiversity conservation and sustainable use, it is crucial to ensure sufficient legal space within seed laws and intellectual property rights for farrmers to continue saving, using, exchanging and selling farm-saved seeds and propagating material. It is also important to consider that farm-saved seed includes both local traditional varieties and improved varieties that have been further adapted and developed by farmers.

To ensure a broad genetic basis of crops, both seed laws and plant breeders' rights must contain exceptions allowing small-scale/local farmers to save, exchange, use and sell farm-saved seeds to other small-scale/local farmers, as long as it takes place in local markets and among local farmers (see Box 9.2).

The definition of what constitutes a 'local' market is complex, and it must take into consideration not only administrative/political and agronomical aspects, but also sociocultural ones. Some proposals aimed at balancing intellectual property rights and farmers' rights to save, use, exchange and sell farm-saved seeds (of protected varieties) have already been suggested by different stakeholders, for example limiting the farmers' rights to save, use, exchange and sell seeds (of protected varieties) to crops produced for consumption at the national level (i.e. this right would not apply to export crops) or limiting the aforementioned farmers' rights to crops used for human food or animal feed (i.e. this right would not apply, for instance, to ornamental plants). Both proposals are legally and politically feasible, and should be considered by countries that are parties to the FAO International Treaty on Plant Genetic Resources for Food and Agriculture, when they implement farmers' rights at the national level.

Box 9.2. The role of 'curadoras' in the conservation of quinoa varieties in the Mapuche communities in southern Chile.

Max Thomet and Didier Bazile

The quinoa (*Chenopodium quinoa* Willd.) was domesticated about 7,000 years ago in the Andes, where it has remained a staple of the indigenous populations of Colombia, Ecuador, Peru, Bolivia and Chile. This very rustic crop has adapted to a diverse range of environments. Although often associated with the Altiplano where it is mainly grown by the Aymaras Indians, quinoa is also part of the farming history of the Mapuche Indians of southern Chile. Archaeological evidence suggests a recent domestication, between 1,200 and 1,400 AD. Once a dietary staple, its consumption is low today. It is still used, however, to make *Mudaï*, a drink consumed during religious ceremonies. In its Inca or endemic varieties, the Mapuche quinoa, also called *Kinwa* or *Dawe*, has therefore never completely disappeared in spite of the introduction of many exotic species during the Spanish conquest.

Tukun is the Mapuche concept of the small vegetable garden (*la huerta*) which the natives formerly farmed. It reflects the cosmogony of the Mapuche people in its shape, its placement and the type of plants grown, the use of the lunar calendar for planting and transplanting, and the oral transmission of knowledge. The Mapuche vegetable garden was earlier located within the forest as a microcosm of food in perfect landscape mimicry. Even today, the responsibility

Box 9.2. Continued.

of the vegetable garden lies with the womenfolk; taking care of plants – as of babies – is considered a feminine speciality. Thus, for thousands of years, seeds were collected, exchanged and adapted in semi-controlled spaces. Self-production and seed exchange is one of the mechanisms that helps maintain and even increase diversity from year to year. As observed in various traditional and complicated exchange systems elsewhere in the world, some farmers in Mapuche communities participate more frequently than others in exchanges. They thus have a wider diversity of species and varieties in their fields than do the rest of the community. These agrobiodiversity 'resource persons' assume the role of the group's '*curadoras*'. Not only are they the seed keepers and protectors of plants, they are also the repositories of the genetic resources and the associated food, medicinal and other knowledge. They also have the responsibility of maintaining the transmission chain for these resources and this knowledge so that the varieties endure over time.

In Mapuche communities in southern Chile, 85% of the quinoa is still grown by women who each sow, on average, between 2 and 4 indigenous varieties. There currently exist two types of exchanges which form the main access to seeds. The first is an 'individual' exchange (between persons or families) within the community based on trust and undertaken in secret. The second is a kind of traditional market (*Trafkintü* in *Mapudungun*, the Mapuche language) organized in the form of public events (in the sense that participation is open to various communities) with an introduction ceremony and the presentation of each participant. The '*curadoras*' correspond to nodes in a seed exchange network and hence control both types of exchanges.

Today, these women '*curadoras*' participate actively as instructors in activities relating to biodiversity conservation. In particular, they work with the ANAMURI (*Asociación Nacional de Mujeres Rurales e Indígenas*) to defend indigenous seeds within the framework of the '*Semillas campesinas: Patrimonio de los pueblos al servicio de la Humanidad*' national campaign. They participate in events that are held daily in various parts of the country, events that are based on their communities' own methods such as fairs and seed exchanges (*Trafkintü*). They take part in seminars and public awareness days. In recent years, they have focused all their efforts on preventing the adoption of international conventions and treaties such as the 1991 act of UPOV (International Union for the Protection of New Varieties of Plants) and the related Chilean national laws because they restrict indigenous rights to free access to and exchange of seeds.

In conclusion, taking into account and benefitting from this knowledge of managing and preserving agrobiodiversity by traditional agricultural communities remains fundamental. Any new strategy developed to preserve biodiversity *in situ* needs to incorporate this knowledge and these practices to reverse ongoing processes of genetic erosion hastened by conventional agriculture and so-called modern seed systems, private or state-owned (Aleman *et al.*, 2010).

9.4.3 Agrobiodiversity and cultural heritage law

To Carl Sauer (1986), cultivated plants are 'cultural artifacts,' and to Laure Emperaire et al. (2008), they are 'biological objects in nature, but cultural in essence'. Culture is even present in the term 'agriculture', and the word 'culture', historically, has also meant cultivation of the land. Culture and agriculture are, therefore, intimately related, and we can use legal instruments aimed at safeguarding cultural heritage to recognize and promote agrobiodiversity-rich agricultural systems and all their elements, both tangible (such as agroecosystems and cultivated plants) and intangible (agricultural techniques, practices, and knowledge).

Safeguarding traditional foodways and dietary diversity is also an important way to promote agrobiodiversity and food security. The conservation of crop genetic diversity cannot be dissociated from crop use, and promoting traditional diets, which are generally more diverse, safeguards humanity's biological and cultural heritage at the same time. Therefore, cultural heritage legal instruments – such as the UNESCO (United Nations Educational, Scientific and Cultural Organization) Convention for the Protection of the World Cultural and Natural Heritage and the UNESCO Convention for the Safeguarding of the Intangible Cultural Heritage – can be used (and in some cases, have been used) to promote agrobiodiversity and food diversity in different and innovative ways.

Historically, UNESCO's Representative List of Intangible Cultural Heritage has focused primarily on performing arts and crafts. More recently, however, UNESCO seems to have accepted that culinary traditions are a cultural expression as fundamental to identity and worthy of recognition as dance, theater or music. As food heritage is directly associated with crop genetic diversity, it is worth mentioning that in 2010 three culinary systems were included in the list: the traditional Mexican cuisine; the Mediterranean diet of Spain, Greece, Italy and Morocco; and the gastronomic meal of the French. UNESCO is also partnering with Bioversity International and the Kenyan Ministry for National Heritage and Culture on a project called 'Safeguarding Traditional Foodways of Two Communities in Kenya'. The project recognizes that traditional foodways, both those related to everyday life as well as those associated with special occasions (such as rituals and festive events), constitute an important part of the intangible heritage of local communities.

Another initiative in the same direction took place in Brazil: on November 8, 2010, the Brazilian federal agency for cultural heritage protection (IPHAN) recognized, for the first time, an agrobiodiversity-rich traditional agricultural system as 'intangible cultural heritage of Brazil': the traditional agricultural system of the Negro River region, in the northwest of the Brazilian Amazon. According to Laure Emperaire, Lúcia Van Velthem and Ana Gita de Oliveira (Emperaire *et al.*, 2008), in the context of the Negro River, an agricultural system may be understood as 'a whole set of knowledge, myths, oral expressions, practices, products, techniques, artifacts and other associated manifestations and expressions which involve the management of spaces, the cultivation of plants, the transformation of agricultural products and local food systems.' These authors explain that the notion of system links cultural heritage (to which recognition is sought) to a wider and more complex set of social relationships. To Emperaire *et al.* (2008), the Registry of the Negro River agricultural system as intangible cultural heritage is 'a concrete example of how instruments and policies for safeguarding cultural heritage can be used in favor of agrobiodiversity, cultural diversity and local agricultural systems' (Box 9.3).

Box 9.3. From phytogenetic resource to cultural heritage: the social bases of agrobiodiversity management in Central Amazonia.
Ludivine Eloy and Laure Emperaire

The Amazon region can be described as megadiverse due to its spontaneous biological diversity and that of its cultivated plants which are carefully selected and maintained by local populations. The future of this agrobiodiversity is now being threatened by the spread of so-called modern agricultural models and techniques, expanding infrastructure, urbanization, etc. Traditional seed systems are being depleted, both in terms of phytogenetic resources at stakes as well as of knowledge and practices associated with them.

The Local Communities, Agrobiodiversity and Associated Traditional Knowledge (PACTA) programme, conducted in cooperation by Unicamp and IDR and coordinated by Mauro Almeida and Laure Emperaire, explores the future of this low-input agriculture and the agrobiodiversity associated with it.

Research undertaken in Brazilian Amazonia, in particular in the Negro River region, is aimed at understanding the social and cultural bases of such systems. The originality of the approach is to focus not on the knowledge as represented by the corpus of operational data but rather to try to understand its social significance and the values that underpin it. We show that the range of diversity managed by the indigenous people far exceeds the functional, agronomic or food requirements. Analysis of seed exchange networks between farmers reveals two levels of agrobiodiversity management. The first operates within the farm: it is the result of slash and burn agriculture, which requires an annual transfer of germplasm from the old swidden field to the newly cleared one. The second network is formed at a regional scale and sometimes extends hundreds of kilometres. The issue is not only the exchange of new varieties: exchanged plants are also the vectors of social significance. Possession of a large number of species or cultivated varieties brings with it prestige and social standing. The diversity of the statuses of cultivated plants within society is revealed by the form this circulation takes: gift, exchange or transmission. These local conceptual models clash with – or complement – the knowledge formalized in agriculture development institutions. The issue is to identify how these local strategies of managing biodiversity can be recognized and valorised in terms of their contribution to the conservation of phytogenetic resources (Emperaire *et al.*, 2010; Cardoso *et al.*, 2010).

On December 23, 2009, Peru issued National Directorial Resolution (Resolución Directoral Nacional) No. 1986/INC, declaring, as 'national cultural heritage', the knowledge, practices and technologies associated with the traditional cultivation of maize in the Sacred Valley of Incas, in the Peruvian Andes. This was also the first time that an agricultural system was recognized as cultural heritage in Peru. According to this resolution, maize or Sara (in the Quechua language), despite being a cereal of Meso-American origin, has been cultivated in the Peruvian Andes for millennia, and 55 maize varieties can be found in this region.

It is also worth mentioning that, in 1992, the UNESCO Convention on the Protection of the World Cultural and Natural Heritage (known as the World Heritage Convention), established a new and innovative category of protected site: the 'cultural landscape'. 'Cultural landscapes' are justifiably included in the UNESCO World Heritage List when interactions between people and their environment are evaluated as being of 'outstanding universal value'. Some important agricultural landscapes have been inscribed under the World Heritage List as 'cultural landscapes', such as: the rice terraces of the Philippine Cordilleras; the archaeological landscape of the first coffee plantations in the southeast of Cuba; the agricultural landscape of southern Öland, in Sweden; and the Puszta pastoral landscape of the Hortobagy National Park, in Hungary. These examples show the possibility of using the 'cultural landscape' category to promote and safeguard agrobiodiversity heritage sites (Singh and Varaprasad, 2008). The rich genetic heritage of crops and livestock associated with diverse agroecosystems, as well as agricultural techniques, practices and innovations held by local communities, can be protected not only through environmental law instruments but also via cultural heritage safeguarding tools, such as the ones mentioned above.

9.4.4 Agrobiodiversity and protected areas

The role of protected areas in promoting the conservation and sustainable use of agrobiodiversity has been greatly underestimated. At the international level, there is still no specific category of protected area for agrobiodiversity conservation, whether for *in situ* conservation of crop wild relatives or for on-farm management of agrobiodiversity-rich farming systems (though other categories of protected areas have occasionally been used for agrobiodiversity conservation).

The number of protected areas in the world has grown from approximately 56,000 in 1996 to about 70,000 in 2007, and the total area covered has expanded in the same period from 13 to 17.5 million km^2, but areas with the richest agrobiodiversity (such as centres of origin and/or diversity) have received significantly less protection than the global average, according to the Second Report on the State of the World's Plant Genetic Resources for Food and Agriculture (FAO, 2010: 34). This report adds that a significant number of plant genetic resources for food and agriculture (PGRFA), including crop wild relatives and useful plants collected from the wild, occur outside protected areas (in cultivated fields, orchards, grasslands, etc.) and consequently do not receive any form of legal protection.

As several authors have already pointed out (Phillips, 2005; Rössler, 2005; Santilli, 2005, 2012), there has been a shift in the conservation paradigm, from designating exceptional natural sites without people (pristine or near-pristine areas) to recognizing the value of natural heritage sites in a landscape context that includes people. That is, more emphasis has been placed on human – nature interaction, and the importance of protected areas that focus on lived-in landscapes has been increasingly recognized. Following this trend, an agrobiodiversity reserve (or heritage site) or a protected 'agrobiodiverse landscape' category could be created, focusing specifically on the management of agricultural biodiversity and on the recognition of traditional agricultural techniques, practices and knowledge. Most initiatives to conserve agrobiodiversity inside protected areas have taken place in parks, biological reserves, etc., which were not conceived for the conservation of crops and cultivated areas. Some exceptions are the Parque de la Papa, created by Indigenous communities of Peru, and the Sierra de Manantlán Reserve in Mexico.

The creation of an internationally recognized protected area category especially aimed at conserving agrobiodiversity would help to promote public awareness about the importance of agricultural diversity and its implications for food security, nutrition, health, social equity and environmental sustainability. Historically, most conservation efforts and policies have focused on wild biodiversity, as if conserving wild plants and animals were more important to humanity than conserving the diversity of agricultural crops, such as rice, beans, maize, potatoes, which are part of our day-to-day diets. In addition, the lack of an integrated approach to agricultural and environmental policies leads, in many cases, to separate and unarticulated efforts to conserve threatened wild species and ecosystems by environmental agencies, and *ex situ* conservation of domesticated crops by agricultural research institutions, without much interaction between them.

The creation of agrobiodiversity reserves would also draw more attention from policy-makers to the relevance of identifying 'agrobiodiversity hotspots', surveying and inventorying areas with high agricultural diversity (especially crop wild relatives, traditional/local landraces, and ingenious farming systems), and defining criteria and methods for the identification and adaptive management of such areas. Through these initiatives, scientific understanding of on-farm management of agrobiodiversity would increase, as would the recognition of the value of local seed systems and social networks in maintaining plant genetic diversity. A category of protected area especially designed to promote on-farm management of agrobiodiversity should allow sustainable use of plant genetic resources and focus on the interaction between humans and nature and on adaptive management.

As it is traditional and local agricultural systems that conserve and manage most of agricultural diversity, an agrobiodiversity reserve could be created only with the support and involvement of farmers who live in the designated area. The involvement of farmers would also be essential in the development and implementation of management plans and actions, which should promote an integrated (and complementary) approach to scientific knowledge and traditional knowledge systems. Agrobiodiversity reserves must meet broader objectives of sustainable local development and social inclusion, rather than just environmental conservation, in order to be socially and politically sustainable, especially in developing countries.

Agrobiodiversity reserves need not necessarily be established in publicly owned lands, and it does not make sense, evidently, to expropriate the lands of farmers within their limits. Agrobiodiversity reserves could be established through agreements with local farmers who live in the designated areas and manage agrobiodiversity-rich agroecosystems. They could be compensated for the conservation of crop diversity through the payment of environmental services, for instance.

9.4.5 Agrobiodiversity and Globally Important Ingenious Agricultural Heritage Systems

In 2002, FAO started an initiative for the conservation and adaptive management of Globally Important Agricultural Heritage Systems (GIAHS). The initiative aims to establish the basis for international recognition, dynamic conservation and adaptive management of GIAHS and their agricultural biodiversity, knowledge systems, food and livelihood security, and cultures throughout the world. GIAHS are defined as 'remarkable land use systems and landscapes which are rich in globally significant biological diversity evolving from the co-adaptation of a community with its environmentand its needs and aspirations for sustainable development.' The GIAHS program is supported by the Global Environment Facility (GEF, through the UNDP). Other partners include UNESCO, the International Center for the Study of the Preservation and Restoration of Cultural Property (ICCROM), the United Nations University (UNU), the International Fund for Agricultural Development (IFAD), Bioversity International and IUCN (International Union for Conservation of Nature). Universities, the private sector and civil society organizations also participate in pilot projects.

The concept of GIAHS is distinct from, and more complex than, a conventional heritage site or a protected landscape. Even though the GIAHS project recognizes that the experience of UNESCO with the identification of World Heritage Sites and landscapes, particularly the category of continuing, organically evolved landscapes, has been useful to promote agrobiodiversity, it considers that such category would need to be complemented by a more agricultural focus and a land-use system approach. The possibility of including agricultural systems under existing cultural landscape categories or eventually creating a new category of World Heritage is being explored by the GIAHS project, which aims to establish an internationally accepted system for the recognition of GIAHS. The GIAHS project expects to facilitate mainstreaming of agrobiodiversity conservation in national biodiversity policies and plans to improve the capacity of countries (where pilot systems were established) to promote sustainable use of agrobiodiversity. It intends to benefit mainly local traditional family farming communities and Indigenous peoples. The project aims to create a GIAHS network and a long-term open-ended programme. Approximately 200 agricultural systems have been identified, and pilot systems have been established in the following countries:

- The Andean agricultural system (Peru), in the valleys of Cuzco and Puno, in the proximity of the Inca city of Machu Picchu. Agricultural crops are divided in terraces, which reach 4,000 meters in altitude, and the region is the center of origin for potatoes (domesticated by the Aymara and Quechua Indigenous peoples), quinoa, cinchona, coca, amaranth, chili and roots of great importance in regional diets, such as arracacha and yacón.
- The agricultural system in the Chiloé Archipelago, in southern Chile, is a centre of origin and diversity for potatoes, and approximately 200 native potato varieties are still cultivated by the Huilliche Indigenous people, as well as a garlic variety (*ajo chilote*) that exists only in the volcanic soils of the Chiloé Archipelago.
- The rice terraces in the province of Ifugao, in the Philippine mountains, which have also been recognized as a World Heritage Site by UNESCO under the 'cultural landscape' category of the World Heritage Convention. It is an agroecosystem of high mountains, in which terraces interact with a set of micro-basins, which function as irrigation and filtering systems. The rice terraces are located along the curves of the mountains. It is estimated that 565 rice varieties are conserved (Nozawa *et al.*, 2008).
- Agricultural systems in the oases of the Magreb region (Algeria, Morocco and Tunisia). The oases of the Magreb region are green islands flourishing in a constraining and harsh environment. They are home to a diversified and highly intensive and productive system which has been developed over millennia. The Tamegroute oasis, in Morocco, also participates in the UNESCO program 'Man and the Biosphere' and is part of the 'biosphere reserve of southern Morocco oases'. It is a highly diversified agricultural system, which is intensive and productive, developed over thousands of years. These areas produce dates, other fruits (pomegranates, figs, peaches, apples, grapes, etc.), legumes and vegetables, cereals, medicinal plants, etc.
- Rice-fish agriculture in China. Fish are raised in flooded rice fields, in an ecologic symbiosis: fish provide fertilizer for rice, regulate micro-climatic conditions, soften the soil, disturb the water, and eat larvae and weeds in the flooded fields, and rice provides shade and food for fish. It is a very traditional agricultural system, which has existed since the Han Dynasty, 2000 years ago.
- Hani rice terraces system, in China. These rice terraces are located in the Honghe Hani and Yi Autonomous Prefecture, in the southeast part of Yunnan Province. The typical ecological landscape of the Hani rice terraces is composed of the forest, village, terrace and river. Hani minorities have lived in this landscape for over 1300 years.
- Wannian traditional rice culture system, in China. Wannian traditional rice is a remarkable old and prototypical variety. It is a variety unique to Heqiao village, and it can be grown only in the water, soil and climatic conditions that prevail in Heqiao village. The traditional rice needs the perennially cold spring water for irrigation, and surrounding forests play a crucial role in soil and water conservation. The surrounding forests and paddy fields are part of the same biodiversity-rich agroforestry system.

- Pastoral system and upland agroecosystem (Kenya, Tanzania). Over 75 percent of the African population lives in rural small-scale holdings and family farms. The aim of this project is to enhance the viability of smallholding and traditional agriculture and agropastoral systems and enhance food and nutritional security of Indigenous communities depending on these systems in Kenya and Tanzania. The project intends to address adaptive management and conservation of the productive landscape of the Masaai and Tapade communities.

The GIAHS programme aims to identify, define and support forms of conservation and dynamic management of these agricultural systems, in order to allow local farmers to maintain biological diversity and at the same time conserve the natural resources necessary for their survival. It seeks to develop public policies and incentives for *in situ*/on-farm conservation of biodiversity and associated traditional knowledge. One of the characteristics of these traditional agricultural systems is precisely their wealth of agrobiodiversity: at least 177 distinct varieties of potatoes have been identified in the Peruvian Andes region, approximately 20 traditional rice varieties are found in fish-rice farming in China, and over 10 distinct varieties of dates have been identified in the Magreb oases. In addition to acknowledging the value and protecting traditional/local agricultural systems, the GIAHS programme can provide inputs for discussions regarding creation of protected areas especially aimed at conservation of agrobiodiversity.

9.5 Conclusion

The world's agrarian and agricultural universe is extremely complex, due to the existence of not only a huge diversity of landscapes and ecosystems (with different environmental features), but also of different types of farmers, who have different livelihoods and develop different survival and production strategies. The reference to a 'duality of agricultural models' (industrial agribusiness versus small-scale farming) is commonplace, but, in reality, our agricultural biological and cultural diversity is not comprised of a duality, but rather of a multiplicity of farming systems (indigenous, traditional, family, peasant, organic, as well as different types of monocultures). The legal framework for agrobiodiversity is still under construction, internationally and in most countries, which gives all social stakeholders a valuable opportunity to reflect and discuss how policies and legislation can contribute more effectively to the *in situ* and on-farm management and sustainable use of agrobiodiversity, as well as to food security, cultural diversity and social equity. Many legal instruments discussed in this chapter are still in the process of being elaborated and/or are being revised. More instruments still need to be constructed and conceived, while others were developed at the international level but must be implemented at the national level. This provides a rich opportunity for law and policymakers to innovate, and to create new tools aimed at strenghtening the complex biological, social and cultural processes that sustain and enrich agrobiodiversity.

References

Aleman, J., Thomet, M., Bazile, D. and Pham, J.-L., 2010. Central role of nodal farmers in seed exchanges for biodiversity dynamics example of 'Curadoras' for the quinoa conservation in mapuche communities in south Chile. In: Innovation and Sustainable Development in Agriculture and Food – ISDA 2010, Montpellier, France. Available at: http://hal.archives-ouvertes.fr/hal-00530950/fr/.

Bioversity International, 2010. Agriculture, agricultural biodiversity and sustainability. 14th meeting of the Convention on Biological Diversity's Subsidiary Body on Scientific, Technical and Technological Advice, Nairobi, Kenya.

Bonneuil, C., Demeulenaere, E., Thomas, F., Joly, P.B., Allaire, G. and Goldringer, I., 2006. Innover autrement ? La recherche face à l'avènement d'un nouveau régime de production et de régulation des savoirs en génétique végétale. In: Gasselin, P. and Clèment, O. (eds.) Quelles variétés et semences pour des agricultures paysannes durables? Les Dossiers de L'Environnement de l'INRA, 30, pp. 29-52.

Cardoso, T., Eloy Pereira, L. and Emperaire, L., 2010. Role des dynamiques spatio-temporelles dans la conservation de l'agrobiodiversite des systemes agricoles amerindiens du bas rio Negro (Amazonas, Bresil). In: Innovation and Sustainable Development in Agriculture and Food – ISDA 2010, Montpellier, France. Available at: http://hal.archives-ouvertes.fr/hal-00512235/fr.

Cromwell, E., Cooper, D. and Mulvany, P., 2003. Defining agricultural biodiversity. In: Conservation and sustainable use of agricultural biodiversity: a sourcebook, vol. 1. Centro Internacional de la Papa (CIP) and Users' Perspective with Agricultural Research and Development (UPWARD), Manila, Phillipines, pp. 1-12.

Emperaire, L., Velthem, L.H.V. and Oliveira, A.G., 2008. Patrimônio cultural imaterial e sistema agrícola: o manejo da diversidade agrícola no médio Rio Negro (AM). Presented at the 26th Meeting of the Brazilian Anthropology Association, held between 1-4 June 2008, Porto Seguro, Bahia, Brazil.

Emperaire, L., Almeida, M., Carneiro Da Cunha, M. and Eloy, L., 2010. Innover, transmettre. La diversité agricole en Amazonie brésilienne. In: Innovation and Sustainable Development in Agriculture and Food – ISDA 2010, Montpellier, France. Avaialable at; http://hal.archives-ouvertes.fr/hal-00512260.

FAO, 1996. Rome declaration on world food security and world food summit plan of action. World Food Summit, November 13-17, 1996. FAO, Rome, Italy.

FAO, 2010. Second report on the state of the world's plant genetic resources for food and agriculture. Commission on Genetic Resources for Food and Agriculture, FAO, Rome, Italy. Available at: http://www.fao.org/agriculture/crops/core-themes/theme/seeds-pgr/sow/en.

GTZ (German Technical Cooperation), 2006. Agrobiodiversity – the key to food security. Sector project 'People, Food and Biodiversity' (Division 45). GTZ, Eschborn, Germany.

Hocde, H., Carlos Rosas, J. and Araya, R., 2010. Co-desarrollo de variedades entre agricultores, científicos y profesionales, biodiversidad y otras cosas. In: Innovation and Sustainable Development in Agriculture and Food – ISDA 2010, Montpellier, France. Available at: http://hal.archives-ouvertes.fr/hal-00531488/fr/.

Louwaars, N., 2007. Seeds of confusion: the impact of policies on seed systems. PhD dissertation, Wageningen University, Wageningen, the Netherlands.

Nozawa, C., Malingan, M., Plantilla, A. and Ong, J., 2008. 'Evolving culture, evolving landscapes: the Philippine rice terraces', in Amend, T., Brown, J., Kothari, A., Philipps, A. and Solton, S. (eds.) Protected Landscapes and Agrobiodiversity Values (Values of Protected Landscapes and Seascapes, a series published by the Protected Landscapes Task Force of IUCN's World Commission on Protected Areas), IUCN, Gland, Switzerland.

Phillips, A., 2005. Landscape as a meeting ground: category V protected landscapes/seascapes and world heritage cultural landscapes. In: Brown, J., Mitchell, N. and Beresford, M. (eds.) The Protected Landscape Approach: Linking Nature, Culture and Community. IUCN, Gland, Switzerland, pp. 19-35.

Rössler, M., 2005. World heritage cultural landscapes: a global perspective. In: Brown, J., Mitchell, N. and Beresford, M. (eds.) The Protected Landscape Approach: Linking Nature, Culture and Community. IUCN, Gland, Switzerland, pp. 37-46.

Santilli, J., 2005. Socioambientalismo e novos direitos: proteção jurídica à diversidade biológica e cultural. Editora Peiropolis, ISA/IEB, São Paulo, Brazil.

Santilli, J., 2012. Agrobiodiversity and the law: regulating genetic resources, food security and cultural diversity. Earthscan, London, UK.

Sauer, C., 1986. As plantas cultivadas na América do Sul tropical. In: Ribeiro, B. (ed.) Suma Etnológica Brasileira: Etnobiologia, 3rd ed. Vozes, FINEP, Petrópolis, Brazil, pp. 59-90.

Singh, A. and Varaprasad, K.S., 2008. Criteria for identification and assessment of agrobiodivesity heritage sites: evolving sustainable agriculture. Current Science, 94(9), 1131-1138.

Chapter 10. Policies to foster innovation in the Mediterranean region

Karim Hussein and Khalid El Harizi

10.1 Context: global and regional challenges for agriculture and food – the innovation imperative

It is now widely accepted that innovation in science, technology and farming practices is essential to address the challenges faced by food systems, particularly in low and middle income countries. The critical role of smallholder and family producers that produce most food in developing countries, along with that of efficiently functioning agricultural commodity chains, is also increasingly being recognized at all levels of policy, action and research. This also implies developing innovative partnerships among actors all along commodity and value chains.

Innovation in agriculture and food, with special attention to smallholder farmers, has climbed to the top of the international and development policy agendas. Following the food security initiatives of the Group of Eight developed countries in recent years[1], innovation, research and development of agricultural productivity were highlighted as priorities by Ministers of the Group of 20 important industrialized and developing countries at the Montpellier G20 Conference on Agricultural Research for Development in September 2011, and re-emphasized by G20 leaders at their Summit in Cannes, France, in November 2011. They underlined the need to foster innovation-sharing with and among developing countries. They also placed a high priority on addressing food security and volatile food prices through investments in science and technology to improve productivity and knowledge-sharing.

This heightened global concern is the result of a combination of factors. First, concern that rapid population growth will increase the demand for food more quickly than agricultural productivity can increase. Indeed, UN population monitoring has indicated that the global population reached 7 billion in October 2011 and is projected to go beyond 9 billion by 2050. To feed this world population, the G20 have estimated that agricultural production will have to increase by 70% globally over the same period. Second, recent severe global food price volatility and price spikes raise concerns over adequate supply of food. Food price spikes in 2008 and 2010 led to instability and deaths in some developing and middle income countries (including West and North Africa). The trend is set to continue for key cereals and staples on

[1] See, for example, the agriculture and food security topics of the UNECA/OECD (2010/2011) Mutual Review of Development Effectiveness in Africa, which detail G8 and G20 commitments to support African agricultural development and food security and reviews the extent to which these have been delivered (http://www.africapartnershipforum.org).

international and regional markets, predicted by the FAO and the OECD to remain some 20% higher in the 2011-2020 period than over the 2001-2010 period (OECD-FAO, 2011). Third, the predicted negative effects of climate change and climatic variability on agriculture and food production and productivity which will affect arid and semi-arid countries the most, such as those in North Africa, the Sahel and the Mediterranean region.

To these global concerns can be added the enhanced competition in globalized and regional markets that require efficiency and productivity gains; soaring energy prices that increase the price of key agricultural inputs (such as fertilizer); and situations where there is instability, conflict, weak institutions, inequality and problems of governance. Together, these constitute key challenges for sustainable development in agriculture and food in the 21st century, requiring innovative responses in the agriculture and food sectors of developing and middle income countries in particular.

As 'best practices' become obsolete or require adaptation to changing environments, development actors constantly need to develop effective new responses. In order to have a lasting impact on rural poverty, these complex and multi-faceted challenges need to be addressed and opportunities seized; this requires the ability to innovate.

Innovation has a key role in enhancing the opportunities and competitiveness of rural people and the agriculture and food sectors in an increasingly globalized world. Case studies from Africa have shown that certain conditions must be fulfilled for access to such innovation to be enhanced: use of participatory approaches; partnerships between agricultural producers and their organizations, agricultural service providers, government bodies and private sector actors; fostering synergies between the formal and informal sources of knowledge and innovation; access to finance and credit; an institutional context that fosters learning and innovation; and global networks that promote dialogue among innovators (Hussein, 2001; Juma, 2010). These require a positive enabling environment supported by appropriate and innovative policies. Innovation and *innovative policies* have indeed been identified by the UN as at the heart of achieving the UN Millennium Development Goals and reducing poverty in Africa in particular (UN, 2010).

Policies to support innovation, targeted research and effective knowledge-sharing and agricultural extension mechanisms are not enough on their own. Appropriate policies provide an essential foundation to increase the productivity, profitability and environmental sustainability of agriculture and food systems – especially for smallholder producers. However, these need to be backed by increased resources, public and private, national and global, to generate effective agricultural transformation that benefits all poorer smallholder producers and consumers, in North Africa in particular. Given the dramatic economic, social and political challenges faced in the North African part of the Mediterranean region, this has become a policy imperative for governments, regional and international organizations. Governments in the region have a vital role and responsibility to provide an appropriate policy framework that encourages and provides equitable access to innovation in agriculture and food.

This chapter reviews global and national policies that foster innovation for sustainable development in agriculture and food, focusing on case studies from the Mediterranean region. In Section 10.2, we begin with a brief review of the importance of innovation in promoting agricultural and rural development and key concepts of innovation.

It draws on practical experience from concrete cases where policies and strategies to promote agricultural innovation have been developed and implemented at the national and regional levels – Morocco and the Near East and North Africa region – challenges experienced and lessons learned. Case study examples include national and sub-national experiences in Morocco, the establishment of a regional network of agricultural research systems across the Near East and North Africa, and development partner work on policies and strategies that promote innovation (IFAD, AFD and the OECD)[2].

10.2 Concepts of innovation and implications for policy, research and action

The concept of innovation is complex and multidimensional and has been addressed in other contributions to this publication and the wider literature on innovation (see, for example, Berdegué, 2005; Biggs, 2008; Hall, 2006; Hall *et al.*, 2001; IFAD, 2007; World Bank, 2006). Here innovation can be seen simply as the search for a better solution to a challenge and one that involves change. Challenges can be understood as 'problems' that need to be addressed and for which effective solutions need to be found. The *quality* of solutions is a key element of innovation.

Innovation can be conceived of as a *process*, not only an output, and one that involves *continuous learning*. A simple model of the innovation process involves the analysis of local circumstances and the recognition of specific problems or issues to be solved, articulation of demand, development of an innovative solution and its testing and implementation in the field. Successful innovations may be disseminated, shared and '*scaled up*' by involving a wider number of actors and '*scaled out*' by implementing the innovation in different contexts. The process of innovation can occur in an incremental or a radical way. In the first instance,

[2] IFAD: International Fund for Agricultural Development; AFD: French Development Agency; OECD: Organization for Economic Cooperation and Development.

innovations might consist of small improvements that together could make a difference in people's lives with a low degree of uncertainty. Radical innovation involves the development of a new solution with a higher degree of risk and uncertainty, given that it modifies current practices.[3]

Partnerships among researchers, universities, agricultural service providers, the private sector (including producers) and civil society organizations favour the development of more effective and efficient solutions to specific challenges to sustainable agricultural and rural development. A supportive environment and policy context that both fosters new partnerships and provides a framework for diverse actors to work together effectively to encourage technological, institutional and policy innovations is indispensable.

Thus, when discussing innovation in agriculture and food for sustainable development, the *policy dimension* needs to be considered carefully. The role that national policies play, and that regional and multilateral organizations have in promoting innovation, collaboration and cooperation to support innovative processes, requires special attention.

Three core questions need to be considered when addressing policy issues in relation to innovation in this area:
1. What are the main drivers and spaces of agricultural and rural innovation for sustainable development at the national, regional and global levels?
2. What institutions have been successful in mediating and promoting learning and innovation and what support do they need?
3. What governance frameworks are required to allocate resources and manage complexity effectively for innovation (e.g. participatory learning approaches, public policies, financing, and assessing impact)?

The ISDA Policy Roundtable in Montpellier in 2010 helped to identify some key elements of successful policies to promote innovation in agriculture and food that are currently being implemented in the region. It provided the basis for this chapter.

10.3 International frameworks and strategies to foster innovation in science and technology[4]

Agriculture and food security policies are extremely important issues for global development, recognized at the highest level by leaders of the G20 given the extreme food price volatility

[3] Innovative techniques are often new practices in a certain local context, rather than the generation of completely new ideas, although their development could be the result of local creativity and innovation rather than simple technology transfer (Berdegué, 2005).

[4] This section draws on a presentation by Mr. Gang Zhang, Principle Administrator, Science, Technology and Industry Directorate, OECD, Paris, France.

in global and regional markets in 2008 and 2010 that affected key staple foods, as noted at the start of this chapter.

A recent study by the Organization for Economic Cooperation and Development (OECD) has indicated that the doubling of prices for agricultural produce has primarily not been due to market speculation; and the OECD-FAO Agriculture Outlook predicts that the prices for farm commodities are likely to stay high and rise in the coming decade – though perhaps not reaching the peaks of 2007-2008, during which prices of key staple grains increased four-fold. Hence, the agriculture and food security challenge is unfolding at the global level and it is very timely to discuss how to address this by fostering innovation. In fact, the OECD has increasingly worked on promoting innovation, food security and agriculture over recent years, including the organization of a meeting of experts to discuss how agriculture can increasingly become a knowledge-intensive sector.

It is now widely recognized that innovation must be part of the solution to agricultural development, productivity and food security challenges. Key global bodies and international organizations have therefore developed *international frameworks and strategies* to foster innovation in science, technology and agriculture. In this context, the OECD developed its innovation strategy in 2010 (OECD, 2010). Recognizing that there is no 'one-size-fits-all' innovation policy, the OECD Innovation Strategy recommended a set of *innovation policy principles*, which can be applied in OECD countries as well as in developing countries.

These principles emphasize:
- empowering people to innovate, through education and skills training;
- putting in place conducive policy and financing framework conditions to unleash innovation, including competition frameworks, making financing available, establishing intellectual property rights, etc.;
- creating and applying knowledge, through investment in R&D, facilitating knowledge flow and use through the fostering of clusters and networks;
- applying innovation to address global and social challenges through improved international scientific and technological cooperation and technology transfer, including through the development of international mechanisms to finance innovation and share costs of addressing global challenges, including food security and agriculture;
- improving the governance and measurement of policies for innovation, including ensuring policy coherence by treating innovation as a central component of government policy, with strong leadership at the highest political levels.

The recommendation of principles is not always sufficient, and countries need further help with implementation. The OECD assists countries in improving their National Innovation Strategies through OECD country reviews of innovation policy. These reviews provide an assessment of innovation performance in a country, and make policy recommendations on how to improve the national innovation system. The OECD has undertaken reviews not only of OECD countries, but also of Chile, China, and some emerging economies, such as South Africa. With regard to developing countries, it is now undertaking one for the South East Asia region, on Vietnam (jointly with the World Bank) and on Peru (with the Inter-American Development Bank).

The OECD jointly published a study with the International Development Research Centre, Canada, in August 2010 that explores the 'Innovation and the Development Agenda' in more detail (OECD/IDRC, 2010). This study examines the role of innovation in developing countries, with a focus on Africa. It emphasizes that innovation drives long-term economic growth and examines innovation systems and their application, the key role of knowledge in innovation for development and the importance of comparable country studies and official statistics on innovation. It stresses the need for innovation to become part of a comprehensive development agenda, and makes recommendations for promoting activities in both the formal and informal sectors, with the aim of transforming agriculture into a knowledge-based industry capable of stimulating economic growth.

In addition, the OECD works on governance for multilateral cooperation in science, technology and innovation to address global challenges, which include the areas of agriculture and food security. This project is needed to improve the contribution of science, technology and innovation in addressing global challenges, such as climate change, agriculture and food security, energy security and global health. The project will address the governance issue by carrying out analysis of the need for new approaches and by making recommendations concerning new governance mechanisms. It focuses on four key governance issues for innovation policies: (1) agenda and priority setting; (2) funding mechanisms; (3) options for institutional arrangements; and (4) turning science into 'real' solutions. This work will propose recommendations for OECD principles and best practices in the area of innovation policies and strategies, as appropriate.

Box 10.1. Example of an inter-regional network to foster agricultural innovation: the regional Association of Agricultural Research Institutions in the Near East and North Africa (AARINENA).[5]

The experience of this *inter-regional network* provides useful lessons for cooperation and strategies needed at the regional level to foster agricultural innovation, bringing insight to three guiding questions.

1. Institutional drivers and spaces of agricultural and rural innovation for sustainable development at the regional level: the role of agricultural research.

[5] Section based on a presentation by Professor Osama Momtaz, Coordinator for AARINENA Network Agricultural Biotechnology Network / Deputy Director for Research, AGERI-ARC, Egypt (see http://www.*aarinena*.org).

Box 10.1. Continued.

National strategies that target smallholders require some adaptation of research and innovation systems. At the same time, national strategies for research need to target strategic national priorities in the sector so that they foster innovation in order to respond to wider challenges. This also implies the need to strengthen national infrastructure and human resource capacities to assess sustainable agricultural development. Linking up with international agricultural research and development communities through regional platforms for coordinated policy and learning is of critical importance for National Agricultural Research Systems (NARS) – as they have common interests and all have to adapt to changing international norms and regulations (e.g. the constantly evolving regulations for biosafety, an area where increasing regional cooperation is developing). Strengthening regional cooperation in capacity building and human resource management through mutual learning across NARS in the Near East and North Africa region is an important area where regional networks such as AARINENA can provide added value.

2. Support needed for research institutions to promote learning and innovation: the case of National Agricultural Research Systems (NARS) in the AARINENA Network.

NARS remain key players in developing and promoting agricultural innovation. These were built to serve national needs and to assess agricultural research development in learning, innovation and research. Yet now they need to engage and fit with the regional and global technical dimensions of agricultural and rural development, as well as policy management, in order to influence and interact with the regional and global levels. For this they need to develop closer linkages with universities and practitioners.

3. Institutions and governance frameworks required to allocate resources and manage them effectively to foster innovation systems that support sustainable agricultural development that also responds to the needs of the poor.

Sound national policies to promote the development and dissemination of agricultural technology, to enforce appropriate regulations, to establish effective management and risk management strategies (e.g. systematizing risk assessments and cost-benefit analysis) and to develop communication strategies contribute to the further development and application of agricultural research and innovation for sustainable development. These national agricultural technology and innovation development strategies should be prepared within the context of the overall development strategy of the country and enforced by policy makers.

While there are similarities across country contexts and the challenges faced, it has proven difficult in the past to achieve the necessary cooperation between countries to promote agricultural and rural innovation for sustainable development in the Mediterranean region. In the context of the rapid social and political changes underway in North Africa, once national priorities for reform have been addressed it can be hoped that more attention will be given to improving the regional cooperation required to achieve common goals with regard to food security and agriculture.

A wide range of countries can benefit from joining international and regional initiatives aimed at addressing shared challenges facing agriculture (for example Box 10.1). The formulation of a national innovation policy needs to take into account international and regional cooperation, which in turn should be part of the national policy for innovation.

10.4 National strategies and policies to foster agricultural innovation in the Mediterranean region: Morocco's new 'Green Plan' agricultural strategy[6]

Rural areas and the agricultural sector have seen major transformations over the last 50 years in Morocco. This sector constitutes about 19% of national GDP, employs 4 million rural people on 1.5 million farms, with some 13.5 million rural people largely dependent on agriculture for their livelihoods, and supports the food security of up to 30 million consumers in the country (Ministère de l'Agriculture, Morocco, 2009; Ministère de l'Agriculture, Morocco, 2008). However, serious limitations to progress have emerged in recent years. These are in part due to the difficulties experienced in adjusting the country's agricultural sector to new economic, social and environmental challenges. They are also due to changes related to globalization, and the opportunities and constraints this process has created. Agriculture has experienced difficulty in fully adjusting to these changes to remain competitive and optimize its contribution to national economic growth. The challenges faced obliged Morocco to react and to innovate, by investing better in the motors of progress and getting the best out of the country's potential and its human resources.

The Government considers agriculture to be the main motor of national economic growth for the current decade, while recognizing the social stakes for both producers and consumers, and the need for agricultural development to be sustainable, taking account of water scarcity and the need for rational territorial development. Thus social, agricultural and economic innovation is at the heart of Morocco's agricultural strategy – the Green Morocco Plan. This national agricultural development plan, launched in 2008, is seen as an *innovative national development strategy* to stimulate investment in agricultural development in Morocco, taking account of the economic and social importance of the sector. The plan has two core pillars: first, the development of high value-added and high productivity agricultural production (targeting some 400,000 farmers with up to 900 projects); and second, social development and poverty reduction (targeting some 600,000 to 800,000 farmers with 400 projects). To implement these interventions, the Plan aims to attract some 150-170 billion Moroccan dirham of investment into the sector over ten years (2011-2020). This case study illustrates how countries in the region hope to optimize the contribution of the agricultural sector to the overall development and national GDP in the Mediterranean region.

---

[6] This section draws on the presentation by Dr. Mohamed Ait-Kadi, President of the General Council for Agricultural Development/CGDA, Morocco.

The Green Morocco Plan represents a major break with the dominant development model of earlier years, which treated 'efficient modern' agriculture and marginalized 'traditional' agriculture as opposing approaches to development. This former model of agricultural development and growth projected a limited vision of agricultural development, inhibiting the strong potential for innovation given the wealth and diversity of agriculture, and involved very limited mobilization of actors involved in the sector. Minimal attention was paid to taking advantage of market opportunities. In this way, it marginalized small farms.

The plan is based on a real change of paradigm. It is built around *two core principles*: (1) 'agriculture for all' and (2) fostering a diversified agriculture that is both socially inclusive and 'modern'. It is based on the following foundations: first, that agriculture will be the main source of growth for the next 10-15 years; second, that if producers join together in an organization in partnership with other actors such as the State, finance institutions, input providers and agro-industry, then poorer, marginal producers can benefit from better profit margins and incomes; third, that agriculture is for all – both high value-added large-scale irrigated agriculture and smaller-scale community or family agriculture in the mountainous, and oasis zones; fourth, the need to attract private investment to complement state financing; fifth, a contractual approach to the implementation of up to 1,500 concrete projects; sixth, safeguarding natural resources to ensure agriculture is sustainable; and seventh, restructuring the sectoral framework (reforming land, water and fiscal policies; modernizing markets and improving access to markets; and restructuring the Ministry to better accompany, monitor and evaluate change).

To respond to the challenges and objectives for the sector identified above in an inclusive way, national and sub-national agricultural policies and strategies need to prioritize the development of agricultural producer capacities, acknowledging that innovation is at the heart of this approach and strategy. Here, fostering innovation includes: taking account of the diversity dimension; reforming agricultural research and development systems by fostering a better interface between farmers, producers and researchers, and establishing regional agriculture clusters (or '*agropôles*') that bring together agricultural research and other agricultural services to support marketing (including product and marketing services and private-sector operators); creating resource centres for innovation dedicated to small-scale farmers, that can act as mediators or 'innovation brokers'; signing of regional plans between national and local government while involving all stakeholders. Various reforms are needed in order to implement the plan.

The Green Morocco Plan was developed to provide a concrete solution to real challenges and represents a major break with the development model of the last 50 years. A new consultative, negotiated and contractual approach has been taken to implement the strategy through the agreement of negotiated production and price targets for specific value chains. This has been implemented *vertically* by commodity supply chain, involving contracts for the development of priority commodities, negotiated with representatives of professional/

producer organizations in the subsector, and *horizontally* at the level of actors involved in local agricultural and territorial development, through regional and local contracts agreed with actors involved in the value chain and regional/local authorities. These two types of activities have been supported by horizontal, cross-cutting initiatives relating to land and water management, marketing and trade policies, financing and R&D.

The following lessons for effective national strategies and policies to foster agricultural innovation in the region can be drawn from this approach:

- the importance of transactional approaches to clarify and bring together the roles of the State and different types of actors/producers engaged in the sector in a policy making process;
- the importance of decentralization: distributive governance and local democracy being embraced in policies to promote innovation, so that local leaders with knowledge of local innovations emerge and take part in the policy and strategy process;
- the need to foster institutional innovation;
- financing must be available to support innovation and policy implementation (including from the private sector/banks); and
- that real progress in this area must be based on a strong entrepreneurial spirit.

10.5 AFD-supported programmes to foster agricultural innovation at the local level in Morocco: incentives, governance and implementation modalities[7]

Micro-local level aspects of innovation processes also need to be taken into account in national innovation policies, specifically organizational and institutional innovation. Examples from development programmes implemented in Morocco with the support of the Agence Française de Développement (French Development Agency), among others, provide useful insights on lessons that can be drawn from local initiatives and local institutional dynamics for the formulation and implementation of national legal frameworks, such as the Green Morocco Plan. A review of project experience by ENFI, IAV, ENA, University of Rabat, IRD, AFD, CIRAD[8], and others working on water services, agro-pastoral services and other aspects of the water and agriculture sectors in Morocco in the 1990s, are instructive for national policies to promote innovation. These experiences involved different approaches, but have by and large been positive in mobilizing actors and improving local natural resource management techniques.

[7] This section draws on a presentation by Claude Torre at the ISDA Round Table, France (Chargé de Mission, Agence Française de Développement, Rome, Italy).

[8] ENFI: National School of Forestry Engineers, Morocco; IAV: Hassan II Agronomy and Veterinary Institute, Morocco; ENA: National Administration School of Rabat, Morocco; IRD: Institute of Development Research; AFD: French Development Agency; CIRAD: Centre for International Cooperation in Agronomic Research for Development.

The 'Haut Commissariat au Plan' (HCP – Prospective 2030-2007) underlined the limitations of former agriculture policies in Morocco, due to an excessively interventionist and technocratic approach that did not involve sufficient ownership by rural actors. Institutional innovation, building on local autonomy and institutions, and building sustainable institutions jointly with local actors ('co-construction') have been shown to be key. Similarly, the evaluation of several programs of integrated rural development supported by the World Bank concluded that continuity of financial support was necessary and that this required effective decentralization of powers and responsibilities.

The collaborative management of natural resources (CMNR) study, a summary report of lessons learned from different rural development projects, showed that these programmes have focused on traditional forms of CMNR. The best known of these is the Agdal (a common grazing enclosure). A programme was implemented that aimed at modernizing these, by incorporating the following elements: written formalization of collective rules; legalization of local development associations that have functional autonomy; establishment of ethno-lineage cooperatives, where the elements of democracy and active participation of youth and women are introduced; territory charters; contracts for use of wood and pastoral resources; and systems of payment for environmental services, etc.

Key lessons from these experiences include the following:

- It is important to be aware of, to understand and to integrate these local institutions, rights and rules with a national law or policy to promote sustainable agricultural and rural development and innovation into a kind of 'hybrid' legal framework that reflects local realities, that links national policies with local plans. *Co-construction* of policies that respect local autonomy is a critical success factor for a sustainable and effective national innovation strategy or policy.
- Local actors and institutions are the key drivers of a successful policy to promote innovation in agriculture and food. Local institutions and producer organizations/cooperatives should be legally recognized in development policies and plans affecting a geographic area. Their inclusion also implies a need to build the capacities of local actors to engage and shape policy.
- A spatial or territorial approach to implementing a national rural development and innovation policy facilitates empowerment of and coordination with the various players: local organizations; civil society; agricultural service providers, etc.
- To engage with and strengthen local institutions takes time: it requires a long-term incremental approach to policy development and implementation. National strategies need to be able to adapt to and incorporate the diversity of actors and their needs. Decentralization must be effective, allowing local governments to mobilize financial resources to support these local institutions.
- Innovations come through the development of 'hybrid laws' and the culture of contracts, which implement negotiated economic and sustainable incentives.

- National innovation strategies and plans remain important in any case to foster the linkage between local and national level development and innovation in the sector with challenges arising from the global economic, environmental and trade context.

10.6 Implications from the case studies for policies to foster innovation in the Mediterranean region

The case studies analyzed above demonstrate the importance of policies that are designed to foster innovation in sustainable development of agriculture and food. They also yield a number of lessons concerning the successful development and application of pertinent policies in this area.

Policies to foster innovation in the Mediterranean region need to take account of political, social and geographical realities: first, diversity of contexts even within one 'region' like the Mediterranean area. Indeed there are two very distinct parts of the Mediterranean region, the northern and southern side, with very different contexts, levels of economic development and resources, constraints and opportunities.

Water scarcity, land limitations, food deficit and conflict over land use are all common challenges that require innovation across the region. There is a need to focus on what can be done jointly to address such common challenges.

In developing national policies to promote a radical change with the past to address emerging challenges in sustainable development in agriculture and food, it is important to devise ways

Box 10.2. A dual process of learning from an institutional innovation: the case of participatory irrigation management in Morocco.
Zakaria Kadiri, Marcel Kuper, Mostafa Errahj

In 1995, a national conference on the new international doctrine of 'participatory irrigation management' was jointly organized in Marrakesh by the Moroccan Ministry of Agriculture and the World Bank. Its purpose was to insert this doctrine into national policies designed to encourage the withdrawal of the State from irrigation management by involving local water users' associations. It was around this institutional innovation that the State established two irrigated schemes in the Moroccan Prerif basin: Middle Sebou in 1994 spread over 6,500 ha and Sahla in 2005 covering 3,242 ha. A second phase of the Middle Sebou project is currently under construction to increase the area covered by an additional 4,600 ha.

Due to its technical configuration, originally designed for centralized management by the State, the Middle Sebou scheme experienced the same problems that large gravity-fed schemes did: water shared by several associations, difficulty of recovering charges for the water used cycle, complicated hydraulic equipment (pumping stations, mainly) requiring regular major

Box 10.2. Continued.

maintenance and technical expertise not always available locally. The involvement of farmers through their water users' associations occurred only gradually, often after the introduction of the scheme. Despite this late participation, water users' associations gradually took over the management of the scheme. They recruited local young people to manage the network's operations. They devised new rules for managing water supply cycles to facilitate night irrigation and alleviated peak demand by encouraging riverain users to irrigate by private pumping. They also introduced rules to facilitate recovery of water dues, for example, by asking users to pay usage charges before the water is supplied to them. They also created a single point of reference by appointing a technical director, thus preventing potential demands by some associations' office holders for preferential supply. Some associations have included new fields into the scheme or have obtained direct access to irrigation water.

The Sahla project was largely based on the experience of Middle Sebou, in part because the team responsible for setting up the scheme had been earlier involved in implementing the Middle Sebou project. In contrast to the latter, the Sahla team was located on-site and consisted of leaders, young graduates of the area, responsible for encouraging and supporting the establishment of water users' associations. Recognizing that agricultural considerations were not sufficiently taken into account during the construction of the Middle Sebou scheme, the team set up a programme of trainings and meetings to facilitate the creation of professional agricultural organizations, not just limited to water users' associations.

Examination of the Middle Sebou-Sahla model as an institutional innovation shows a dual learning process: by the farmers and by the administration. Many water users' associations have gradually become increasingly effective actors though, it must be admitted, some have failed to do so successfully. The appropriation of institutional innovation begins with the associations mastering the management of the scheme by suitably adapting the rules. It then continues with agricultural leaders becoming involved and active in other spheres, such as Agricultural Product Boards (APB), local development or even local politics. For the administration, this innovation has been a real laboratory: it is no longer apprehensive in delegating the management of a new irrigation scheme to farmers, and State agents have become accustomed to being in constant negotiations and interactions with farmers.

Finally, our discussion reveals that institutional innovation is not limited to a 'technical component' of managing irrigated schemes. It depends also on a political will at the national level to 'democratize' these schemes and on power relationships between various interest groups on the ground. As far as the transition is concerned, the implementation of participatory irrigation management is not without difficulties since it competes with other models of action. National policies consider and explore other innovations, some of which are presented as more effective. This applies, in particular, to private-public partnerships for managing irrigated schemes. The place of farmers in such partnerships is often unclear at first glance. Will water users' associations emerge as full partners in such mechanisms or will they be marginalized in the interactions between the administration and private companies? (Kadiri et al., 2010).

to avoid disrupting existing resource access and use rights and to take specific risks regarding women and their rights into account.

Successful policies need to take account of local experience (see Box 10.2). Capacity building of actors in the sector – both those in government and at the local level – is critical to successful policies. Experience in capacity building in the sector reveals that the main capacity lacking is the ability for staff to work effectively at the local, field level. This human capacity is difficult to acquire, and formal training is not sufficient. In doing this, however, it is important that the State does not squeeze out local and farmer initiatives. Innovation results from the interaction between policy and actors: a change of mindset, vision and mentalities. Innovation goes beyond participation and needs negotiation between interests and actors.

As has been vividly demonstrated by the dramatic changes of the 'Arab Spring' of 2011, government bureaucracies and institutions in the region, including those involved in agricultural research, development and policy, need to reform the way they operate and set their priorities in order to respond to the hopes and expectations of citizens – whether they are rural or urban. But the degree to which they can reform effectively to satisfy citizen demands remains unclear.

Research and development systems in Mediterranean countries, particularly those in North Africa, need to be reshaped to address a radical and participatory innovation agenda, guarantee equitable use of natural resources, develop appropriate responses to climate change challenges in agriculture and learn from change in both the resource base and people's behaviour in order to remain relevant. There is a need for research and development institutions to increase their capacity to work closely with development and local institutions. Also, research needs the strong support and engagement of the State and sufficient resources from all stakeholders to generate pertinent and useful results (see Box 10.3).

However, policymakers often consider the concept of 'innovation' as too loose or general for application to national policies for sustainable agricultural and rural development. Indeed, more clarity is needed on the appropriate role of government in fostering innovation in agriculture and food. In addition, as the policy environment is constantly changing, this

Box 10.3. Institutional innovations to help adoption of technical innovations for cereal cultivation in Tunisia.
Khaldi Raoudha

The surge in global cereal prices in 2007 led to the doubling of the cereal purchase bill for Tunisia. This led the country to develop a new cereal strategy (2009-2013) based on the promotion of irrigated cereals, as an alternative technique to increase yields, which have

Box 10.3. Continued.

remained low and unpredictable. This *political* choice has to be accompanied by changes at the institutional level for an improved transfer of technical innovations and a better research-extension-profession relationship.

The new political strategy advocates three main institutional innovations:

- *Creation, in 2009, of a National Institute of Cereal Crops (INGC)* whose role is to enhance the use and application of scientific knowledge. The emphasis is, primarily, on applied research, communication with farmers and extension. The cereal farmers will benefit from the institution's expertise and know-how, either directly or through professional organizations, development groups and agricultural services cooperatives. The approach is based on programme contracts.
- *Setting up federated research projects (FRP) in the domain of cereal crops.* To better fulfil its coordinating role, the Institute of Agricultural Research and Higher Education (IRESA) has put in place, since 1998, various Commissions for Planning and Evaluating Agricultural Research (CPERA). These commissions bring together representatives from the world of research, public administration and the profession (farmers, agribusinesses, distributors). Research is structured around multidisciplinary and multi-institutional Federated Projects (FP) on previously agreed priority themes. The goal is to put an end to top-down, *ad hoc* and isolated interventions and to promote R&D to better meet farmer requirements. R&D agreements have been entered into with the development organizations (Regional Agricultural Development Commissions, development agencies and inter-professional groups). In the domain of cereal crops, a new generation of projects, called Federated Research Projects (FRP), were launched for the 2009-2012 period to address the shortcomings of the initial projects: prioritization of research by major areas, opening of teams to local socio-economic partners (producers, NGOs, policy makers, regional institutions, agribusinesses), networking and assistance in innovation transfers.
- *Involving the profession in innovation transfer efforts.* The Tunisian Agriculture and Fisheries Union, a trade union organization, has, since 2008, strengthened its links with research by launching a unique programme to support farmers through partnership agreements.

In conclusion, the transition to an efficient institutional system is still far from easy. Recent institutional innovations could improve the technical performance of farms but efforts should be directed towards a true participatory approach and a close and targeted extension system. Although research is increasingly better planned, the process of promoting, developing and sharing innovations among actors still remains the responsibility of several institutions, rendering coordination and the drawing up of a coherent plan very complex. The transition planned by the new political strategy needs to be accompanied by the introduction of 'agricultural advisors', representing the new form of private extension to farmers. Such advisors have existed since 1998 but mainly served the bigger producers. To allow a more broadbased action, the setting up of producer groups – of the agricultural development group type – at the local level is encouraged to support advisory services and to establish partnerships for promoting access to agricultural innovation which today only benefits the larger producers (Khaldi *et al.*, 2010).

process of change needs to be taken into account in concepts and approaches to fostering innovation, so it can be managed and adapted to over time.

Local development in the sector cannot be fostered in a way that is disconnected from regional and global development challenges and policies. But there is no one-size-fits-all: the most appropriate policies to address specific problems need to be developed.

Policies in the sector in this region are often too disconnected from local realities and priorities. National visions and policies need to adapt to diversity, local contexts, consumers, markets and available infrastructure – in short, be made more 'human' and coherent. How this change in policy making can be achieved is a major challenge.

The agriculture and food sectors face a fast changing environment. Mediterranean countries need to enhance their capacities in policy analysis to be able to understand and formulate response to such change. It would be useful to establish a regional 'think tank' for the Mediterranean region, creating space and capacity for real dialogue, knowledge-sharing and capacity building across the region – across North African Mediterranean countries, and linking up with European Mediterranean countries.

10.7 Conclusions

Appropriate policy frameworks are indispensable to foster innovation and equitable access to innovation. All the main international development organizations now recognize the vital importance of innovation and have developed their own innovation strategies, e.g. OECD, IFAD and the World Bank. However, innovation in agriculture and food means different things to different people. Family farmers and poor producers and consumers are often not the central focus of innovations strategies and policies. Policies developed to promote innovation need to be flexible and adapted to context, different actors – including poor smallholder farmers – and to the social, economic and political transformations they are undergoing.

Policies to promote innovation in agriculture and food in the Mediterranean region, and North Africa in particular, have been the focus of this chapter. This region has become increasingly significant in 2011 with the dramatic changes in a number of countries – political, economic and social. Hope has mingled with apprehension and disappointments as social movements emerged in many North African Mediterranean countries and demanded radical change, representation and equity.

The G8 countries have launched concrete initiatives and made significant resources available for economic development and governance reform in these countries, for example through the G8 Deauville Partnership between the G8 and four Middle East and North African countries – including Morocco, Egypt and Tunisia. The mandate of the European Bank for

Reconstruction and Development has also been extended to cover North Africa for which massive resources are being made available to support governance reform and investment. This has changed the political landscape to accord priority to equity, employment, and economic and social inclusion in the region in national, regional and international policies. As agriculture remains a key sector of the economies in the region, and food-price stability is key to maintaining urban and rural peace, the importance of developing effective policies that promote agricultural innovation could not be more significant.

From the above we draw the conclusions below, organized according to the three questions posed in Section 10.2. Indeed, these three questions could help guide the periodic review of innovation strategies and policies in the Mediterranean region in the context of a rapidly changing political, social and economic environment in the region.

10.7.1 Drivers and spaces of agricultural and rural innovation for sustainable development

Clarifying the concept of innovation. An innovation is a new and better solution to challenges or obstacles that exist for specific actors in a concrete context. Local actors, private sector, communities and States all have roles in identifying and promoting innovation. In a developing-country context, it is useful to think of innovation as converting knowledge, formal or informal and from whichever source, to added value. Innovation should not be equated with pure invention; innovation is to a large extent a process that begins after invention. Finally, innovation implies a change of mindset, a change or adaptation of former approaches and practices.

Drivers and spaces for innovation. Key challenges such as national or regional food security, food price volatility, environmental pressure and climate change drive the need for innovation. Innovation processes take place in physical spaces (geographical areas, territories, regions, etc.), product value chains, legal contexts, and global environmental and policy contexts. This complex of drivers and spaces needs to be taken into account in the development of successful innovation policy frameworks.

10.7.2 Institutions that have been successful in mediating and promoting learning and innovation and the support they need to be effective

Role of local institutions, value chain organizations and actors. Local actors, private sector entrepreneurs and local initiatives are central to innovation occurring: they are the ones that convert knowledge into value for the economy as a whole. Institutional innovations at the local through to sub-national and national levels are critical to addressing the challenges in the development of the agriculture and food sectors.

Capacity building. Building the capacities of actors at all levels in innovation systems, from the poorest family farmers to those in agribusiness, research and development systems, and government departments dealing with agricultural research and development is vital. This will help to ensure innovations are appropriate and demand-led, while ensuring policy frameworks are adaptable and flexible enough to respond to diverse realities and opportunities.

Policy analysis and peer review. Countries need to enhance their ability to undertake policy analysis at the national and regional levels in order to formulate appropriate responses to emerging challenges, and share lessons across different parts of the country and between countries at the regional level. Networks can be helpful here (such as the AARINENA Network described in Box 10.1). Also, policies to foster innovation need to be built around a deeper understanding of interactions between different stakeholders faced by a given challenge that might be amenable to an innovative response. A think tank for the Mediterranean region would be useful for this purpose, along with developing real capacities for dialogue across the region and between the northern and southern sides of the Mediterranean.

Policy Learning. Peer-learning, cooperative international learning platforms (such as provided by the OECD's technical and policy fora) and South-South learning are needed to improve policies. Recording lessons from case studies and developing appropriate materials or tools to encourage such learning is critical.

10.7.3 Governance frameworks required to allocate resources and manage complexity effectively for innovation

Role of government. The State is not the only or even the lead player in complex innovation processes. However, it does have an important role in creating incentives, an environment that encourages and rewards innovation, and in developing the capability of a society to innovate. Governments need to provide appropriate conditions, resources and incentives to encourage innovation by a range of actors. For this there is a serious need for the reform of the public sector and service providers in the region. This *innovation support role of the State* can be categorized in four clusters: (1) identifying the challenges faced in the country/region that require better solutions; (2) developing people's skills in creativity and innovation management; (3) establishing and helping establish open spaces and processes for learning and innovation, where innovation risks are better regulated and managed, and the likelihood of success is increased and/or accelerated; and (4) creating incentives and eliminating obstacles to innovation. This means placing innovation high on the public policy agenda.

Regional integration and regional platforms. Regional integration, regulated open markets for goods, technology and labour, can promote more rapid innovation processes with good results for productivity and sustainable development, as has been shown in several regions in the world. Regional platforms for sharing approaches, best practices and innovations as well as for developing coordinated policy frameworks to promote innovation are key.

Ensuring social equity, participation and equal access to innovation. Policies to promote innovation need to pay adequate attention to the social and technological dynamics of innovation if they are to succeed in the long term. They must therefore address issues of social equity and access 'up front', and recognize that innovation processes and collective action 'learning processes' can exclude some actors and individuals in communities. This means ensuring that smallholders and family farms have voice and access to innovation, but also that women, young people within households and minorities often living on marginal lands have equal access to opportunities offered by innovations. Innovation systems must be participatory, at all levels of decision-making and implementation, and give voice to all stakeholders.

In summary – the experience presented in this paper highlights seven essential elements of successful policies to foster innovation and sustainable development in agriculture and food:
- Appropriate and complementary policy frameworks and strategies to foster innovation are needed at the national, regional and global levels.
- Policies need to draw on and adapt to actors, contexts, culture and transformation processes. They need to be 'co-constructed' to be effective and relevant.
- Innovation systems must be participatory and include all stakeholders in the concerned societies.
- A wide range of challenges exist for developing relevant policies for sustainable development in agriculture and food, but a general lesson from experience is that they need to be developed, implemented and monitored in a consultative and participatory way.
- For this, it is essential to invest in, and provide incentives for, building capacities – at the local level and also in national research and development institutions.
- The priorities of research and development systems in the sector must evolve and adapt in line with emerging local, national and indeed global challenges and priorities.
- In a changing environment, policy frameworks need to be adaptable and flexible.

References

Berdegué, J., 2005. Pro-poor innovation systems. IFAD, Rome, Italy.

Biggs, S., 2008. Learning from the positive to reduce rural poverty and increase social justice: institutional innovations in agricultural and natural resources research and development. Experimental Agriculture, 44, 39-60.

Hall, A., 2006. Public private sector partnership in a system of agricultural innovations: concepts and challenges. International Journal of Technology Management and Sustainable Development, 5, 3-20.

Hall, A.J., Sivamohan, M.V.K., Clark, N., Taylor, S. and Bockett, G., 2001. Why research partnerships really matter: innovation theory, institutional arrangements and implications for the developing new technology for the poor. World Development, 29 (5), 783-797.

Hussein, K., 2001. Producer organizations and agricultural technology in West Africa: institutions that give farmers a voice. Development, 44, 61-66.

IFAD, 2007. Innovation strategy. IFAD, Rome, Italy.
Juma, C., 2010. The new harvest: agricultural innovation in Africa. Oxford University Press, Oxford, UK.
Kadiri, Z., Belmoumene, K., Kuper, M., Faysse, N., Tozy, M. and Mostafa Errahj, M., 2010. L'innovation institutionnelle dix ans plus tard: quelles opportunités pour les agriculteurs, et quels apprentissages pour les pouvoirs publics? Cas des associations d'irrigants au Nord du Maroc. In: Innovation and Sustainable Development in Agriculture and Food. ISDA 2010, Montpellier, France. Available at: http://hal.archives-ouvertes.fr/hal-00523316/en/.
Khaldi, R., Zied Dhraief, M. and Albouchi L., 2010. Innovations institutionnelles face à la crise pour une meilleure adoption des innovations techniques des céréales irriguées en Tunisie. In: Innovation and Sustainable Development in Agriculture and Food. ISDA 2010, Montpellier, France. Available at: http://hal.archives-ouvertes.fr/hal-00522038.
Ministère de l'Agriculture, Morocco, 2008, Plan Maroc Vert: Premières perspectives sur la stratégie agricole, April 2008.
Ministère de l'Agriculture, Morocco, 2009. Plan Maroc Vert. Powerpoint presentation, May 2009.
OECD, 2010. The OECD Innovation strategy: getting a head start on tomorrow. OECD, Paris, France.
OECD-FAO, 2011. OECD-FAO Agricultural Outlook 2011-2020. OECD, Paris, France. Available at: http://www.agri-outlook.org/.
OECD/IDRC, 2010. Innovation and the development agenda. OECD, Paris, France.
UN, 2010. Assessing progress in africa toward the millennium development goals. UN, New York, NY, USA.
UNECA-OECD, 2010/2011. Mutual review of development effectiveness in Africa. OECD, Paris, France. Available at: http://www.africapartnershipforum.org.
World Bank, 2006. Enhancing agricultural innovation: how to go beyond the strengthening of research systems. Agricultural and Rural Development, World Bank, Washington, DC, USA.

Chapter 11. Designing innovative agriculture policies in Africa

Papa Abdoulaye Seck, Aliou Diagne and Ibrahima Bamba

11.1 Introduction

The year 2010 had a special significance for Africa. For most of its countries, it marked the fiftieth anniversary of their independence. In our view, such a milestone is reason enough to examine the performance of African Agriculture.

Even though the African agricultural sector has grown significantly since the mid-1990s (Badiane, 2008), its performance since the 1960s has been, on the whole, unsatisfactory (World Bank, 2007). Indeed, for some food staples, the African continent, despite its agricultural potential, has experienced a worrying food deficit for several decades.

The African agricultural sector contributes up to 16% of national GDPs compared to an average of about 1 to 8% in other continents. It involves an average of close to 75% of the population but its share of public budgets remains low. According to the World Bank (2007), the share of public expenditure for agriculture was less than 4% in Africa in 2007. Despite the dominance of agriculture as a source of income and livelihood for the majority of the African population, the continent remains a net importer of food staples such as rice and wheat. Thus, Africa is home to 15% of the world's population but accounts for 32% and 25% of world imports of rice and wheat respectively. According to the FAO, nearly one in three Africans suffered from malnutrition in 2006 while this rate is estimated to be 13% for the planet as a whole.

The poor performance of African agriculture cannot be blamed entirely on the lack of suitable agricultural policies but it is partly the result of under-investment in this vital sector and, above all, the lack of relevant and consistent agricultural policies to address the many current and future issues concerning the sector. Rapid population growth and accelerating urbanization constitute two major challenges for African agriculture. Indeed, Africa has recorded an annual population growth rate of 2.5% between 2000 and 2005 compared to 1.2% for the world as a whole. In addition, Africa is urbanizing at a great speed. At present, the urbanization level of the continent is 38%. In 2030, the number of Africans living in cities is estimated to reach 48% (United Nations, 2006). With rapid urbanization and population growth, a major present and future challenge for Africa is to increase agricultural productivity and production in a sustainable manner to be able to feed its populations while preserving its natural resources.

A qualitative transformation of African agriculture is needed to better address this challenge. It will involve a change in the way African agriculture is governed, by moving from a linear and top-down approach of managing the rural economy, where policies are decided by the State without any involvement of the populations concerned, to a decentralized economy, 'co-managed' and 'co-evaluated' by all stakeholders. This revamped method of managing African agriculture requires the development of a stable, coherent and fair environment (political, economic and institutional) that allows various actors of the agricultural value chains to adhere to jointly formulated policy and to build durable and mutually beneficial business relationships. This innovative way of designing agricultural policy is based on a more democratic governance of the sector. It involves acceptance by and effective engagement of all stakeholders in transparent processes for setting priorities and decision making.

In this chapter, we will present some proposals for reforming the governance of African agriculture. We start by drawing up a brief and selective etiology of the under-achievement of African agriculture. We then present some mechanisms and practical measures for bringing about a qualitative transformation of African agriculture.

11.2 Etiology of the poor management of agricultural issues

Increased agricultural production is a necessary condition for feeding the African people, but it is far from being sufficient. Sustainable agriculture development requires consistency in all development policies; agricultural development policies have to be consistent with, amongst others, macroeconomic policies, trade policies, industrialization policies, R&D policies, education and training policies, etc. But in several African countries, agricultural policies were designed in the belief that agricultural performance could be improved in isolation from other aspects of socio-economic life. For example, a large external accounts or balance of payments imbalance or high volatility in exchange rates may undermine policies to revive the agricultural sector. Krueger *et al.* (1988) have shown that the implicit taxation of agriculture resulting from an overvalued exchange rate or from policies for protecting the industrial sector led to agricultural price distortions that characterized the direct taxation of the sector through export taxes.

In the past, African States have tended to focus on productivity approaches which consist of seeking an increase in acreages under cultivation or in yields and thus in overall production. Such productivity approaches exhibit little concern for the viability and profitability of all the links of the agricultural value chain (Badiane, 2008; World Bank, 2007). In other words, the States have tried to address the problems within the farm and ignored those outside. Several signs of this approach's failure are detectable: amongst others, the decline in rural incomes due to the lack of appropriate marketing channels for local production or due to high transaction costs resulting from a lack of infrastructure and poor access to areas of consumption.

Moreover, it was customary for policymakers to determine unilaterally the crops to give priority to, the acreages to sow and the production levels to achieve, all without any consultation with producers, processors and other stakeholders. With this sort of behaviour, which still persists to some extent, the policymakers refuse to recognize that producers and private operators may have useful analytical capacities. In some cases, decision makers order studies which are based on participatory processes but choose not to involve agricultural supply-chain stakeholders in the processes of setting priorities and decision making; they are consulted but there is no co-decision making. To this must be added the inability of State interventions to incorporate innovations which could improve the performance of the value chains. Also of concern are the timid subsidies and their distribution very often on a subjective basis, depending on political considerations.

11.3 Training of actors and poor organization

The training of agricultural value chains actors or the information provided to them is not based on sustainable and sustained programmes of development of new ideas, knowledge and expertise. In general, they are project related and thus of generally short durations. Without appropriate training and information, some skill transfers do not produce the desired results.

Agricultural professional organizations, on their part, suffer from a lack of resources and means. Moreover, they do not adhere to a philosophy and basic principles which would ensure good governance and allow them to defend their professional interests.

Agricultural research is not always perceived as a key sector for fostering sustainable agricultural growth. But studies on the profitability of agricultural research reveal the wisdom of investing public funds in this sub-sector. Rates of return on agricultural research in Africa are generally higher than 30%, sometimes dramatically so (Alston *et al.*, 2000). For example, the mean internal rate of return on investments in research in and dissemination of 'Sahel varieties' of rice in the Senegal River valley is estimated at over 221%, which is well above the cost of access to capital, estimated at 18% for the 1995-2004 period (Fisher *et al.*, 2001). These high rates of return of public agricultural research in Africa justify a call for massive investments in this strategic sector to improve the performance of African agriculture.

Despite promising results, agricultural research does not still find place at the core of the political agenda. Often, policymakers captivated by the yields obtained with some varieties developed abroad order their import without consulting first with national experts. Strategies for transfers of foreign technology have been attempted on many occasions in the past, usually unsuccessfully. For example, most cereal varieties imported into Africa from elsewhere were found to be less resistant to the continent's biotic and abiotic stresses and also to be much more fertilizer dependent. Thus, the direct introduction of sorghum and millet varieties from India into West Africa resulted in very limited success (Matlon and Spencer, 1985). Tellingly, from over 2,000 mangrove rice varieties imported from Asia for tests in African conditions, only two had a performance even comparable to the best local varieties (Matlon and Spencer, 1985).

Results from national agricultural research are often not applied, due partially to the lack of synergy between agricultural research and agricultural extension. Deprived of technical and financial means, agricultural research and agricultural extension often work in isolation. Moreover, the training of actors follows a 'technicist' logic instead of a more comprehensive approach encompassing the technical, economic and commercial aspects. This is one of the consequences of productivity approaches.

11.4 Qualitative transformation of African agriculture

The formulation of innovative agricultural policies in Africa to 'produce more and better' will require a break from the dirigiste, productivity and non-systemic approaches in favour of those that are more participatory and decentralized. To this end, it is necessary to reform the governance of African agriculture, moving from an administered rural economy to a decentralized economy 'co-managed' and 'co-evaluated' by all stakeholders. Also key in this shift is the establishment of a more democratic form of governance involving acceptance by and effective engagement of all stakeholders in transparent processes for setting priorities, decision making and impact assessment. In this chapter, we stress that the qualitative transformation of African agriculture depends on a reform of its governance system and that such a transformation also calls for a broad partnership of all actors in the sector.

To be effective and functional, the reform of the governance of African agriculture must (1) delineate more clearly areas of intervention and operational responsibilities of the various stakeholders while bringing them closer together, (2) promote the specialization of actors according to the principle of comparative advantage, (3) strengthen the synergy and complementarity of actions and interventions of actors operating at the same level of the production chain of the agricultural added value agricultural production chain but in dissimilar disciplinary domains and (4) apply the principle of subsidiarity between actors operating in the same field of activity to maximize the use of limited resources of African States. Going from a consultative approach to a genuinely participatory one will require all actors to be made aware of the requirements of the new approach and to be trained to meet them. In particular, strong and well-structured professional organizations should be promoted so that they can make an active contribution to processes of formulating, implementing and evaluating agricultural policies. In addition, cross-disciplinary training should be offered to producers so that they are better equipped to deal with the diversity and complexity of issues of agricultural development.

11.5 Well trained and informed professional organizations

As in other areas of socio-professional activities, the rural world also needs some balancing forces. Bringing together actors with similar interests in a particular agricultural supply chain is a necessary, but not sufficient, condition. These organizations need to become more professional by adopting a number of principles, amongst them: (1) a shared vision, (2) internal statutes and regulations which are understood and respected, (3) regular meetings by their governing bodies, (4) inclusion of gender-related issues, (5) transparent and auditable accounting, (6) external evaluation of activities, (7) renewal of their governing bodies in accordance with adopted mechanisms and term limits, (8) free flow of information, (9) recognition of the legitimacy of the rights of each member, (10) recognition and respect for individual and collective obligations, (11) adoption of strategic decisions taken by the governing bodies which meet regularly and (12) building up of their professional capacities with the help of actors selected via open tenders when support is State- or project-funded.

Some progress in this regard has been made by African agricultural organizations. An example is the decision taken in 2008 by the Network of Farmers' and Agricultural Producers' Organizations of West Africa (French acronym: ROPPA) to conduct an external evaluation of its programme of activities for the 2000-2008 period. Another ROPPA initiative is to implement a consultative concertation mechanism for priority supply chains (rice, horticulture, livestock/meat, dairy and fishery) as part of an endeavour to improve this association's governance. In addition, the project to federate the African farmer organizations at the continental level is a major step towards an improved structuring of producer organizations. Nevertheless, effective institutional support remains necessary to strengthen the process of professionalization of agricultural organizations. This institutional support must however be defined in consultation with the stakeholders, never decreed or imposed.

Capacity building of professional organizations should not be limited, as is often the case, to production-oriented training. The actors of the agricultural sector have to confront a variety of challenges and the training provided to them should reflect their concerns and expectations. In other words, professional organizations and actors should be better equipped to deal with the complexity of reality, both within and outside their farms.

11.6 Establishment of agricultural value chain observatories

The quality of policies and strategies for agricultural development depends on the relevance and quality of available information and on the quality of mechanisms of interaction between the different actors. Very often, in Africa, information destined for professional organizations and actors is not supplied or updated in a continuous manner. In addition, it is often incomplete because it is limited to the communication of consumer or farm-gate prices recorded in various markets and/or estimates of quantities produced and lists of areas with surplus or deficit productions. Few African countries have the means to establish and maintain market information systems and the creation of agricultural value chain observatories can improve collaboration and interaction between the various actors.

Each value chain observatory should have a framework for exchanges and participatory analysis of problems by the various value chain stakeholders. In an agricultural value chain observatory, representatives of the socio-professional categories are responsible for collecting and disseminating data and information as defined previously by the stakeholders. An observatory should include the following actors at the very least: producers, processors, distributors, consumers, importers/exporters, representatives of State technical services (trade, extension, research, etc.) involved in the operation and regulation of the particular supply chain.

An agricultural value chain observatory can have the objective of developing a partnership and of building mutual trust between the stakeholders for fostering an improvement in the quality of their interactions through information exchanges. The basis of an observatory is the implicit assumption that information users in a supply chain are themselves holders of information. Thus a value chain observatory can provide a framework for collective learning and for strengthening interactions between actors through the establishment of work routines and the development of endogenous mechanisms to overcome challenges or oppositions, even to resolve conflicts. Thus, each observatory participant provides information on the link of the chain that concerns him and receives similar information from the other participants. Collaboration by the supply chain's actors in an observation can help build relationships of trust and lead to the emergence of constructive partnerships between actors of the same value chain. For example, (1) producers can provide farm-gate prices and harvest calendars, (2) distributors can provide wholesale market prices, list of products that are in short supply, retail prices and product quality requirements (3) importers/exporters may provide data on quantities exported by crop, trends in international markets, etc. and (4) the extension services can provide information on varieties used per species, etc. In fact, the observatory's data originates from the information provided by the supply chain's stakeholders. It is more reliable than the traditional information systems because the person providing the information does so in a framework which includes other actors with whom he has interactions and mutually beneficial relationships.

11.7 Governance of agricultural research

The linear and top-down approach of agricultural innovation being generated exclusively in research centres – located at one end of the knowledge-production pipeline – to the adoption of a new product based on that knowledge by passive producers – at the other end of the pipeline – is no longer appropriate. Indeed, there exist a multiplicity of sources of knowledge and technologies. As argued by Hall (2009), innovation stems from the mobilization of collective intelligence in a framework of collaboration between different knowledge sources. To this end, innovation requires knowledge from multiple sources, including from the very users of this knowledge (Hall, 2009). Thus, in the process of adopting new agricultural technologies, producers generally play a role of adapting the new technologies to the specificities of their contexts. Given the complexity and diversity of the environment of poor farmers, technologies that are destined for them require even more effort to be adapted to local conditions (Douthwaite *et al.*, 2001). This applies in particular to smallholder African farmers operating in very heterogeneous environments and production conditions. Therefore, agricultural research undertaken in isolation cannot by itself bring about a qualitative transformation of African agriculture. In fact, the quality and diversity of interactions between farmers, agricultural research, extension and farm advisory services and private sector actors are essential to the process of developing and transforming African agriculture.

Agricultural research also has a role to play in promoting and encouraging interactions and opportunities for interactive learning between the various actors in the agricultural value chain. For example, through its Sub-Saharan Africa Challenge Programme (SSA-CP), the Forum for Agricultural Research in Africa (FARA) facilitates the implementation of a novel approach for promoting innovative processes through continuous interactions between actors, feedback from them, and analysis and absorption of lessons learned from different processes (Hawkins *et al.*, 2009). This approach explicitly recognizes that agricultural innovation does not always start with agricultural research but rather requires an aggregation of the contributions of all actors, including those of the farmers. Agricultural research is a constitutive element of the innovation system and one of its contributions is to meet the demand for knowledge and information, to supplement existing knowledge flows rather than to replace them (Hawkins *et al.*, 2009).

Notwithstanding the enormous benefits of agricultural research and its potential to contribute to growth, this sub-sector is still not a core concern of policymakers. During the 1990s, public investment in agricultural R&D decreased to an average annual rate of 0.2% (Beintema and Stads, 2010). In addition, funding of agricultural R&D in African countries was heavily dependent on foreign donors. Exacerbating the situation is the aging of skilled researchers in national agricultural R&D systems. Retiring personnel have often not been replaced in several countries due to bans on public sector recruitment as part of structural adjustment programmes. Not only does public investment in agricultural research in Africa remain low, it is even in decline in several countries. As a consequence, the scientific capacity in Africa also remains low: the continent contributes only 0.3% of the scientific output in the world and has only 70 researchers per million people against 4,380 for Japan.

The African countries must substantially increase budgetary allocations to agricultural research to maximize its contribution to a qualitative transformation of African agriculture. In this respect, the Global Conference on Agricultural Research for Development (GCARD) held in 2010 recommended that developing countries strive towards endogenous funding of agricultural research by devoting at least 1.5% of their GDPs to this strategic sub-sector. In addition, reform in the way agricultural research in Africa is governed is also necessary. This should take the form of improved coordination and cooperation between national and regional agricultural research systems along principles of subsidiarity and of complementarity and by making use of the comparative advantage of existing institutions. Further, this reform should, on the one hand, promote the effective participation of producers, the private sector and civil society in the governance bodies of research centres and, on the other, redefine relationships between researchers and other professionals within the framework of a new partnership based on consultation and coordination in the planning, implementation, evaluation and valorisation of research results.

11.8 The national agricultural research systems

The responsibility of designing and implementing specific research programmes for agricultural development should be allocated primarily to specialized institutions which have a comparative advantage in the domain concerned. The principle of subsidiarity, both geographical and thematic, must be respected while doing so. Thus, the national agricultural research systems (NARSs) must be in the driving seat of all research activities for agricultural development taking place in their countries, with the support, if necessary, of regional and international research institutions. The support extended by these institutions to the NARSs (which include both national research centres and universities) should be in the context of strengthening national research capacities to enable the NARSs to eventually assume their full responsibilities. Interventions that take recourse to shortcuts by substituting national expertise with regional or international expertise should be banned since they just perpetuate a system of dependence and thus delay the institutional and human development of the NARSs.

11.9 Regional mechanisms for research coordination

At the regional level, coordination and implementation of research programmes and projects in a clearly defined agricultural domain or value chain should be the responsibility primarily of a specialized regional research institution with proven expertise in this area. If no such organization exists, responsibility should devolve to one of the umbrella research organizations, either subregional (WECARD, ASARECA and SACCAR) or regional (FARA). Specialized regional research organizations include intergovernmental organizations and centers of the CGIAR consortium (formely known as the Consultative Group of International Agricultural Research). By definition, a regional research project is one with a purely regional theme or whose implementation or applicability of expected results covers several countries. For example, regional coordination enables the undertaking of activities with a regional scope such as regional exchanges, work in state-of-the-art laboratories, methodological support programmes, etc., with national-level activities (surveys, trials, local exchanges, etc.) handled by national partners. A project not meeting any of these three criteria should devolve to the NARS of the concerned country. In case this latter does not have all the expertise necessary to implement the project, technical support from a regional or international organization or from another country (developed or developing) should be sought within the framework of North-South or South-South cooperation. However, this technical support should be designed with a view of developing local expertise in the concerned domain.

However, to truly fulfil their roles, the regional institutions should go beyond what the NARSs do at a national level. They should do so by technically adding value to the aggregation of the efforts of the NARSs involved. The coordination function alone cannot justify the involvement of a regional institution in the implementation of a research project with a regional scope. The regional institution has to (1) contribute added scientific value to the NARSs involved, (2) undertake a true work of knowledge transfer to and research-capacity building of the NARSs, (3) facilitate learning and sharing of experiences between countries, (4) ensure the quality of research through the organization of critical peer reviews, (5) organize the sharing of research results with countries not involved in the concerned project's execution and (6) undertake the additional work of analyzing, synthesizing and communicating at the regional or continental level. It is obvious that for the contribution of this value addition to be possible and effective, the regional institution has to have a critical mass of researchers with proven technical expertise in the domain under consideration. This requirement is usually satisfied only by regional or continental research centres specialized in specific domains.

In contrast, sub-regional and regional umbrella research organizations (FARA, WECARD, ASARECA and SACCAR) have a comparative advantage in the coordination of agricultural R&D programmes that target multiple agricultural commodities or that relate to cross-disciplinary themes. Such organizations also have a very important role to play in the institutional development of NARSs by promoting the adoption of governance best practices and by helping their management maintain some degree of autonomy from the political environment. Indeed, sudden changes in direction and policies are often decreed by policymakers without genuine consultation with those involved in research, without adequate preparation and without taking into account the priorities set and commitments entered into earlier at the regional and national levels. No organization is better positioned to play this role than FARA and its regional sub-organizations because they have the necessary degree of independence and legitimacy. Improving the governance of the NARSs is a vast undertaking which alone can form a large part of FARA's activities and of its regional sub-organizations for many years to come.

Such a delineation of areas of expertise in the coordination and implementation of research programmes for agricultural development will not only lead to the best use of limited resources but will also help build capacity of national and regional research institutions while reducing duplication of efforts.

11.10 For concerted programming of research priorities

The proposed participatory approach will require an effective participation by the stakeholders in the governance bodies of agricultural research centres, namely the producers, the private sector and civil society. As the main beneficiaries and users of agricultural research, these stakeholders can monitor the definition and implementation of research priorities formalized in a set of specifications. The agricultural world's actors must also participate in the evaluation of research activities and become advocates of research (see Box 11.1).

As for the shared planning of agricultural research, the researchers and users should undertake joint discussions on (1) the characteristics of the value chain in question and the main constraints confronting producers, (2) the formulation of research themes, (3) establishing priorities among the themes, (4) the overall strategy for implementing the research, (5) the potential impact of the research, (6) defining prospective studies and reflections and (7) the outcomes of the research. The exercise can be called successful if the research agenda truly 'belongs to all'. Criticism relating to the relevance of strategic research issues should be viewed as constructive in nature.

The reforms proposed for the governance of agricultural research will require changes in criteria for assessing researchers and their behaviour and attitude towards their environment. Any framework for a shared setting of research priorities should encourage the emergence of a new type of researcher, one who will be able to, on the one hand, decode users' messages and convert them into research themes and, on the other, communicate with different stakeholder categories and better anticipate the social, political, economic and environmental consequences of various events and scenarios. In addition, a 'total researcher' must be assessed not only on his or her scientific productivity but also on (1) the impact of his or her results on the agricultural sector, (2) his or her ability to tackle issues jointly with other specialists, (3) the relationships he or she will be able to develop with national, regional and international partners and (4) the valorisation of his or her research results.

11.11 Rethinking the State's role in the agricultural sector

The State should consider itself as just one stakeholder amongst many others. This means that it should open up the agricultural value chains to other actors through a participatory and interactive approach. In addition to its regulatory function and its mission of equitable redistribution in the public interest, the State[1] must be a facilitator with a full awareness of current and future challenges confronting agriculture and of the sometimes conflicting goals and interests of the various stakeholders. Given that the market in itself is unable to sufficiently stimulate agricultural innovation, the State has a central role in creating a

[1] In its role as guardian of the public interest, the State can define rules (health, social, etc.) binding on all and also legitimately define national priorities.

Box 11.1. *Action research in partnership: a process of reconciliation between research and society.*

Mélanie Blanchard, Eduardo Chia, Mahamoudou Koutou and Eric Vall

The involvement of local actors in research undertaken for rural sector development remains low and their needs and constraints are rarely taken into account when innovations are developed. Consequently, a significant number of research results are adopted either inadequately or not at all by them.

In the agricultural domain, it is only recently that local, on-the-ground actors have participated in defining issues, seeking answers, establishing procedures and then implementing and evaluating solutions. New research approaches, among which Action Research in Partnership (ARP) (Chia, 2004; Faure *et al.*, 2010), are redefining the role and place of agricultural research as well as the relationship between local actors (producers, advisors, technicians of decentralized State services, NGOs) and scientists (Anadon, 2007; Sebillote, 2007; Vall *et al.*, 2008). This original approach was tried out in the Teria and Fertipartenaires action-research projects in western Burkina Faso. These projects, through a partnership built between farmers, technicians and scientists, aimed to improve soil fertility using agricultural innovations co-designed by the partners.

The proposed approach allowed the partners, i.e. the local actors and the scientists, to diagnose the problem, explore possible solutions and undertake actions in consultation with each other, in a spirit of problem-sharing and a willingness to change. This approach was used to co-develop innovations for the production of organic manure (field compost pits, manure pits, cotton stalk composting) and innovative cropping systems to strengthen the role of legumes in the crop rotations (cereal-legume intercropping system, fodder crops, cultivation under cover). Furthermore, a local charter for the management of natural resources was drawn up.

Two major innovations were co-designed within the framework of these projects to render the partnership between local actors and scientists more effective. First is the Village Coordination Committee (VCC), consisting of crop and livestock farmers, women, technical advisors, etc., working under a negotiated ethical framework. The VCC is where decisions are made and information shared and it is the primary interlocutor with the scientists, thus ensuring the active participation of local actors at each research stage (Koutou *et al.*, 2010). The second innovation is the concept of a specifications document or contract that is formalized between the local actors participating in the programme (via the VCC) and the scientists. It specifies each partner's role (who does what) in the experimentation programme (Chia, 2004; Chia and Deffontaines, 1999; Chia *et al.*, 2008, 2009).

Furthermore, rural experiments undertaken within an ARP approach are a source of innovation and learning processes and lead to the production of new local know-how necessary to co-develop innovations (Blanchard *et al.*, 2010). This know-how and learning guarantee a degree of sustainability (Blanchard *et al.*, 2010).

conducive environment for the emergence of diverse types of interactions and opportunities for interactive learning between the different actors of the agricultural value chain (Hall, 2009). Consequently, the State must promote the development of stable institutions and institutional mechanisms for improving the quality of mechanisms and processes of negotiation, interaction, dialogue, co-decision making, co-management and co-evaluation with producer organizations and private operators to enhance the performance of agricultural value chains (Box 11.2).

In addition, it is necessary to always position public interventions and agricultural policies within the overall framework of the strategy for societal development and transformation. The various compartments of socio-economic life have their effects on agriculture and they are not necessarily within the ambit of the ministry in charge of this sector. Examples are market regulation, basic infrastructure, enforcement of agricultural regulations, imports of agricultural products, etc. This means that a transformation of agriculture requires the effective involvement not only of the agricultural sector's actors but also of those working in related fields.

Public services have a key role to play in improving the production environment through appropriate regulatory measures and public investments, as defined in consultation with the stakeholders. On this point, it is necessary to point out that the agricultural sector requires adequate funding. In 2000, in Maputo, the African Heads of States resolved to allocate at least 10% of their national budgets to agriculture. However, six years later, it is clear that only 10 of the 53 African countries have honoured their commitment. In fact, in 2007, the African countries earmarked only an average of 4% of their budgets to agriculture.

It is equally important to improve the quality of public spending on agriculture as it is to increase the quantity. This can be done by aligning the spending with priorities set in consultation with all stakeholders in an objective and transparent manner, by improved training of actors and by strengthening institutions. In addition, improving the quality of public expenditure requires that orientations of and decisions on public investments be based on empirical facts through a rigorous use of scientific results and data. It is worth noting in this regard that subsidies on private goods and services such as fertilizer and credit are less effective than investments in basic public goods such as agricultural research and rural infrastructures (Lopez and Galinato, 2007).

However, given the ever-present possibility of market failures (Omamo and Farrington, 2004), the use of targeted subsidies to certain actors based on objectives of food security, social distribution or the conservation of national resources is economically well-founded. The World Bank (2007) recognizes that subsidies may be justified temporarily to encourage the use of certain seeds and fertilizers even though long-term goals must remain the emergence of viable private-sector based input markets. Africa is the continent with the lowest use of mineral fertilizers in the world with an average consumption of 13 kg/ha in 2008 as compared to 94 kg/ha for all the developing countries combined (Minot and Benson, 2009). A significant

Box 11.2. Innovation platforms enabling innovations in livestock sector in Ethiopia.
Kebebe Ergano

This text box draws lessons on ways to establish and facilitate innovation platforms on smallholder agriculture in developing countries based on experiences from Fodder Adoption Project (FAP) implemented between 2007 and 2010 by the International Livestock Research Institute (ILRI). The project was designed as an 'action research' initiative to address issues in the livestock sector by taking inadequate livestock feeds as an entry point. Its core activities in Ethiopia were to develop and support networks that would diffuse fodder technologies in three pilot districts using innovation systems approaches. One of the clear lessons is that different actors participate in innovation platforms when they see tangible benefits, which justify their investment of time, effort and resources. Tangible benefits, preferably with an early pay-off, were found to be effective in winning trust with farmers and drawing the attention of more stakeholders. Linking forage technologies with a range of value chain issues in dairy and fattening enterprises was essential for successful adoption of forage technologies by farmers. Addressing other overarching constraints in the dairy value chain, such as access to crossbred dairy cows, breeding and veterinary services, milk transportation and market linkages, credit and access to appropriate inputs and information created effective demand for fodder technologies. A diverse network of actors across the entire value chain were needed to turn knowledge into actions and benefits. A wide range of innovation management tasks, i.e. providing access to technology and markets, organizing producers, training, network building, conflict resolution, etc. need to be bundled together to facilitate the uptake of technologies. Our experiences showed that facilitation of the innovation platform is a critical role to turn knowledge into actions and benefits. It needs an individual or a team to take overall responsibility for keeping the momentum going and fulfilling the expectations that the innovation platform raises among different stakeholders. The proponents and facilitators of the innovation platform will need to invest in maintaining the energy and trust among stakeholders as the innovation platform unfolds. However, there was no agency in the districts mandated with the tasks of facilitating innovation platforms and willing to pay for such a service. The task of partnership facilitation among multiple stakeholders (researchers, farmers, traders, processors, policy makers, civic organizations, etc) is not sufficiently recognized by policymakers and there is no incentive to play such a demanding role by the conversional extension workers. The conventional agricultural extension system gives emphasis to supply of technology and inputs. So the FAP project had to fulfil the innovation platform facilitation role and gradually devolved the responsibility to the district agricultural extension system. Therefore, investment in technical research capacity alone will do little to enhance agricultural transformation unless the capacity to broker innovation processes is developed. Capacity needs to be strengthened to manage the collective dynamic of configurations of agencies working in and around a given innovation agenda. A special development fund that can be used for underwriting potentially non-profitable trial investments, covering the costs of workshops and meetings, extra facilities for training and support is needed. Providing competitive grants to the research and university systems for action research that strives for system innovation could go a long way in creating innovation champions. That would also trigger curriculum change and produce a generation of professionals with conviction in innovation systems thinking (Ergano et al., 2010).

increase in crop yields of small farmers is technically unlikely without a policy to promote access to mineral fertilizers. This low level of fertilizer use not only leads to low agricultural productivity but is also responsible for depletion of soil nutrients. Indeed, without the use of fertilizer, the current African crop practices contribute to further soil impoverishment. It is estimated that each year, African soils lose 22 kg/ha of nitrogen, 2.5 kg/ha of phosphate and 15 kg/ha of potassium (Smaling et al., 1997).

Thus to maximize gains of agricultural productivity in Africa it is essential to achieve the twin objectives of increased yields and improved soil quality. This can only happen when producers gain access to mineral fertilizers under favourable conditions. Any policy for promoting the use of mineral fertilizers has to be accompanied by policies for training in the use of organic manure and in the practices of conservation agriculture and agroforestry. However, the means used to achieve these objectives remain subject to controversy. Despite their high fiscal burden, subsidies for agricultural inputs are understandably seeing a resurgence in popularity in Africa. Malawi's experience in substantially increasing its food production – going from a food requirements deficit of 43% in 2005 to a surplus of 53% (Denning et al., 2009) – mainly due to targeted fertilizer and maize seed subsidies has been widely noted. Widespread replication in many African countries of subsidies tailored to the market with instruments such as vouchers, compensatory payments and partial debt guarantees (World Bank, 2007) will go a long way in improving the productivity of African agriculture.

11.12 Conclusion

Defining innovative agricultural policies in Africa is to first accept that breaks from the past will be necessary to 'produce more and better'. Clearly, successful farming is necessarily designed in a framework of co-construction, co-execution and co-evaluation. In fact, each actor of the agricultural supply chains has a contribution to make. And, without doubt, it is the sum of each such individual contribution that leads to new knowledge and processes. It is therefore important for the various actors to maintain a dialogue to be able to establish major innovation systems. Africa has suffered enough from dirigiste, productivity and non-systemic approaches, from the lack of interest in strategic areas such as research and extension, from the marginalization of professional organizations in policy formulation and assessment and from an environment which is unconducive to private sector involvement.

Consequently, a change in mindset is essential and must encompass all actors, public and private. An improved distribution of responsibilities is required and must be the result of a dialogue, dictated by merit, involving all the actors at all stages of the value chain. Africa can and must reject a fatalistic attitude. After all, it has sufficient water and land, large human capital and a number of unused technologies. The continent must reframe its rural development by daring to think differently and invest in its future by embracing a framework of information and cooperative analysis.

References

Alston, J.M., Chan-Kang, C., Marra, M.C., Pardey, P.G. and Wyatt, T.J., 2000. A meta-analysis of rates of return to agricultural R&D. ex pede herculem? IFPRI Research Report 113, Washington, DC, USA.

Anadon, M., 2007. La recherche participative. Multiples regards. Presses de l'Université du Québec, Quebec, Canada.

Badiane, O., 2008. Maintenir et accélérer la reprise de la croissance agricole de l'Afrique dans un contexte de fluctuation des cours mondiaux des denrées alimentaires. IFPRI. Politiques alimentaires en perspective no. 9, Washington, DC, USA.

Beintema, N.M. and Stads, G.J., 2010. Public agricultural R&D investments and capacities in developing countries: Recent evidence for 2000 and beyond. Note prepared for GCARD 2010. IFPRI, Washington, DC, USA. Available at: http://www.ifpri.org/publication/public-agricultural-rd-investments-and-capacities-developing-countries.

Blanchard, M., Vall, E. and Chia E., 2010. Conduire une expérimentation en recherche-action-en-partenariat. Co-concevoir une innovation, l'étudier. In: Innovation and Sustainable Development in Agriculture and Food – ISDA 2010, Montpellier, France. Available at: http://hal.archives-ouvertes.fr/hal-00520255.

Chia, E., 2004. Principes, méthodes de la recherche en partenariat: une proposition pour la traction animale. Revue d'élevage et de médecine vétérinaire des pays tropicaux, 57, 233-240.

Chia, E. and Deffontaines, J.P., 1999. Pratiques et dispositifs de recherches face à un problème d'environnement. Nature Sciences Société, 7(1), 31-41.

Chia, E., Barlet, B., Tomedi, M., Pougmogne, V. and Mikolaseck, O., 2008. Co-construction of a local fish culture system: case study in Western Cameron. Communication, International Farming Systems Association (IFSA) European Symposium, Clermont-Ferrand, France, 6-10 July 2008, 5 p.

Chia, E., Verspiren, R. and Vall, E., 2009. Demande sociale, coproduction de connaissances et émancipation des acteurs. Le cas de la recherche-action-en-partenariat. 2ème Colloque International Francophone sur les Méthodes Qualitatives. Enjeux et stratégies, Lille, 25-26 June 2009, 15 p.

Denning, G., Kabambe, P., Sanchez, P., Malik, A., Flor, R., Harawa, R., Nkhoma, P., Zamba, C., Banda, C., Magombo, C., Keating, M., Wangila, J. and Sachs, J., 2009. Input subsidies to improve smallholder maize productivity in Malawi: toward an African green revolution. PLoS Biology 7(1), e1000023.

Douthwaite, B., Keatinge, J.D.H. and Park, J.R., 2001. Why promising technologies fail: the neglected role of user innovation during adoption. Research Policy, 30, 819-836.

Faure, G., Gasselin, P., Triomphe, B., Temple, L. and Hocde, H., 2010. Innover avec les acteurs du monde rural: la recherche-action en partenariat. Quae, Versailles, France.

Fisher, M., Masters, W.A. and Sidibé, M., 1998. Technical change in senegal's irrigated rice sector: impact assessment under uncertainty.WARDA Ex-Ante impact assessment of rice research bulletin No. 98-1, WARDA, Bouaké, Ivory Coast.

Hall, A., 2009. Challenges to strengthening agricultural innovation systems: where do we go from here? In: Scoones, I., Chambers, R. and Thompsons, J. (eds.) Farmer First Revisited: Farmer-led Innovation for Agricultural Research and Development. Practical Action, Rugby, UK, pp. 30-38.

Hawkins, R., Heemskerk, W., Booth, R., Daane, J., Maatman, A. and Adekunle, A.A., 2009. Integrated Agricultural Research for Development (IAR4D). A Concept Paper for the Forum for Agricultural Research in Africa (FARA) Sub-Saharan Africa Challenge Programme (SSA CP). FARA, Accra, Ghana. Available at: http://www.icra-edu.org/objects/anglolearn/IAR4D_concept_paper.pdf.

Ergano, K., Duncan, A., Adie, A., Tedla, A., Woldewahid, G., Ayele, Z., Berhanu, G. and Alemayehu, N., 2010. Multi-stakeholder platforms strengthening selection and use of fodder options in Ethiopia: lessons and challenges. In: Innovation and Sustainable Development in Agriculture and Food – ISDA 2010, Montpellier, France. Available at: http://hal.archives-ouvertes.fr/hal-00522978.

Koutou, M., Vall, E., Chia, E., Dugué, P., Traoré, L. and Andrieu N., 2010. Implication des acteurs locaux dans la conception d'innovations et l'évaluation de leurs impacts: le cas des systèmes agropastoraux du Tuy (Burkina Faso). In: Innovation and Sustainable Development in Agriculture and Food – ISDA 2010, Montpellier, France. Available at: http://hal.archives-ouvertes.fr/hal-00522574.

Krueger, A., Schiff, M. and Valdes, A., 1998. Agricultural incentives in developing countries: measuring the effect of sectoral and economywide policies. World Bank Economic Review, 2 (3), 255-272.

López, R. and Galinato, G.I., 2007. Should governments stop subsidies to private goods? evidence from rural Latin America. Journal of Public Economics, 91(5-6), 1071-1094.

Matlon, P.J. and Spencer, D., 1985. Increasing food production in Sub-Saharan Africa: environmental problems and inadequate technological solutions. American Journal of Agricultural Economics, 66 (5), 671-676.

Minot, N. and Benson, T., 2009. Fertilizer subsidies in Africa: are vouchers the answer? Issue Brief 60, International Food Policy Research Institute, Washington, DC, USA.

Omamo, S.W. and Farrington, J., 2004. Policy research and African agriculture: time for a dose of reality? ODI Natural Resource Perspectives, No. 90, January 2004. Available at: http://www.odi.org.uk/resources/details.asp?id=2051&title=policy-research-african-agriculture-time-dose-reality.

Sebillote, M., 2007. Quand la recherche participative interpelle le chercheur. In: Anadón, M. (ed.), La recherche participative. Multiples regards. Presses de l'Université du Québec, Quebec, Canada, pp. 49-84.

Smaling, E.M., Nandwa, S.M. and Janssen, B.H., 1997. Soil fertility in Africa is at stake. In: Sanchez, P. and Buresh, R. (eds.), Replenishing soil fertility in Africa. Special Publication no. 51. Soil Science Society of America, Madison, WI, USA, pp. 47-62.

United Nations, 2006. World Urbanization Prospects: The 2005 Revision. UN, Department of Economic and Social Affairs/Population Division, New York, NY, USA.

Vall, E., Chia, E., Andrieu, N. and Bayala, I., 2008. Role of partnership and experimentation for the codesign of sustainable innovations: the case of the West of Burkina Faso. In: Benoît, D. (ed.) Empowerment of the rural actors. A renewal of farming systems perspectives. CD-ROM 8th European IFSA Symposium, 6-10 July 2008, INRA, Paris, France.

World Bank, 2007. World Development Report. Washington, DC, USA.

Chapter 12. Conclusion: en route...but which way?

Bernard Hubert, Emilie Coudel, Oliver T. Coomes, Christophe T. Soulard, Guy Faure and Hubert Devautour[1]

> *Caminante no hay camino, se hace camino al andar*
> Antonio Machado

12.1 Revisiting traditional paths of innovation

Much like the African tradition of the guest telling his host several times that he must 'be on his way' before being allowed to leave, many of us have been saying it is time for us to 'be on our way' along the path to innovation; but a question remains: which way forward now? The ISDA symposium was an opportunity for us to address this issue, to get our bearings, before embarking on our journey.

If the issue of innovation in agriculture is so difficult to address today, it is because there is no one clear objective, no road is signposted. If producing more remains a requirement, it is essential not only to produce more efficiently but also to produce other goods, tangible as well as intangible. Society's demands on agriculture are increasingly complex: from environmental services, inclusion of marginalized populations, and quality differentiation to revitalization of rural territories and energy production, etc. These growing demands challenge us to rethink the role and even the functions of the agricultural sector as well as the place of research in innovation, as pointed out by Lawrence Busch (Chapter 2).

In the context of the modernization of agriculture, promoted as part of the Green Revolution, farmers are expected to maximize production. This requires them to be technically innovative. But increased productivity requires significant investment and incorporation into supply chains, thus reducing the net benefits to the farmer and calling for still greater production: an endless dash forward. This vicious circle of innovation has made farmers question the fundamental basis of their vocation. When adding the negative effects of output maximization on the environment, on food quality, and on the rural economy, this model has become broadly contested in society.

[1] This conclusion is based on the discussions held during the ISDA symposium, in particular during the final roundtable where we gave the floor to several policymakers and civil-society representatives. The authors thus thank the facilitator, Oliver Coomes (McGill University), and the participants of this roundtable: Mohamed Ait-Kadi (General Council of Agricultural Development, Ministry of Agriculture, Morocco), Pascal Bergeret (Ministry of Food, Agriculture and Fisheries, France), Dominique Chardon (Farmer, President of the Association Terroirs & Cultures), Khalid El Harizi (IFAD – International Fund for Agricultural Development), Judith Francis (CTA – Technical Centre for Agricultural and Rural Cooperation), Ann Waters-Bayer (ETC-Prolinnova) and Fabrice Dreyfus (Director of IRC – Institute for Higher Education in Tropical Agri-food Industry and Rural Development).

In response to such concerns, other dimensions of innovation have been gradually incorporated in the effort to advance agriculture: the valorisation of biodiversity, preservation of traditional crops and products; direct sales to revive the links between farmers and consumers; and pluriactivity to help support and maintain agricultural activities, all the while contributing to rural economic growth. Experimentation with new systems often begins with farmers themselves; those who show imagination, creativity and perseverance in challenging conventional ways. But it is clear today that these innovations can only truly take hold, become sustainable over time and offer a viable alternative, if they find support from the broader world of research and encouragement from public policy. Too often, such endeavours are confined to isolated groups of farmers. They manage to make promising breakthroughs over the course of a few years but they then often 'disappear', losing the struggle to mainstream thinking and practice, which remains narrowly focused on labour, land and animal productivity.

Society's attitudes towards research are often contradictory: research is expected to provide the answers to humanity's problems, and society is disillusioned with research advances and rejects its value. For some people in society, the world of research has 'collaborated' with the productivist paradigm; for others, new scientific knowledge and technologies beget new dangers to the environment or to health. Moreover, science now addresses subjects that are so complex that the results may seem confusing even to the informed public. Solutions are not self-evident or self-imposing. We must recreate the subtle link between the objects which research builds and the way they are taken up by society, by defining new relations between science and society. However, these new objects are not only more and more complex, they are also sometimes new – often unknown until now, and build by scientific knowledge. Isn't this the case of climate change, the erosion of biodiversity, transgenetics, emerging diseases, etc.? Isn't this a new paradox for a research in society, of having to explore questions which trouble society? These new questions are often dealt with within the disciplinary fields which generated them, but shouldn't they also be debated within transdisciplinary groups, which would also include more than scientific knowledge?

It is therefore important for research to assume a different role in relation to society; research in which responsibilities are shared, through the development of innovation networks, among scientists, policymakers and actors in civil society. It should be clear to everyone now, thinking beyond our own research specialities, that fully satisfying solutions do not yet exist and that we must now think outside the productivist and linear model, which has long been the main frame of reference. We must define together a new vision. Surely, we cannot respond to future issues using recipes from a model that is questioned by a changing world and new ways of perceiving it. In the 1950s, the vision seemed clear, and a consensus emerged that drove innovation towards the goal of increased productivity, as embodied by the Green Revolution. Research institutions, technical extension institutions and cooperatives were all compatibly organized, working towards this common goal. Today, the path is far from being as clear, in fact, there is no longer a single path; instead, a whole diversity of paths is appearing.

We must give ourselves the means to foster and support this diversity, to strengthen it where it already exists, to manage diversity without impoverishing it, in order to open new options for the future. This task is far more complex because one must identify those paths on which different actors, with differing interests, will agree to advance together for some distance; a journey made even more challenging because it unfolds in an uncertain world. Thus it would seem worthwhile to explore novel paths rather than apply recipes whose desired effects – and unintended consequences – are difficult to control. This is the lesson we have learnt in recent years, as have researchers in many other sectors, such in energy or urban planning.

At the conclusion of the ISDA symposium, we revisited the following issues discussed among participating stakeholders, based on answers found in the experiences and shared by those present:

- How to define a new vision? How to think differently about innovation? How to deconstruct models that hinder new thinking about innovation? How to explore new directions?
- How to achieve that goal? How to conceive the transition to other models of innovation? How to generate change? What knowledge to build?
- How to organize ourselves? How to interact better to create knowledge that can engender innovation?

We present here an initial distillation of the experiences and insights shared at ISDA, hoping that they will stimulate debate during field research on agricultural innovation, within partnership research programmes, and in actors' discussion fora.

12.2 Challenging existing visions of agriculture in order to explore new ones

Our challenge of the hegemony of the productivist model and the type of innovation associated with it is not new. In 1993, an international conference was organized by INRA, Cirad and Orstom (now IRD) on the topic of 'Innovation and Societies: What kinds of agriculture? What kinds of innovation?' The challenge then was to deconstruct the myth of 'technology transfer' and review the limitations of 'top-down' approaches, where both producers and consumers were often reduced to passive actors. Today, the importance of different stakeholders is broadly acknowledged, among NGOs who have become experts in participatory approaches and among multinational companies who emphasize 'social responsibility' and recognize the value of multi-stakeholder approaches. The challenge is no longer to contest and deconstruct the dominant model but to analyze how different visions of agricultural development can co-exist. Whereas most actors agree that sustainable development is a desirable goal, their underlying visions on how to do so vary widely, from the promise of green technologies and the ecological utopia of degrowth to the dream of a unified and united global society.

In considering the co-existence of models, we must not ignore the influence of past trajectories, the phenomenon of 'path-dependency'. If the productivist model remains dominant, it is because the model has proven itself in parts of the world where it was imperative to feed growing populations and avoid civil strife. One of the model's advantages is that it helps rationalize agriculture (by planning, measuring), which is vital for agro-industries and investors. This model finds itself under attack today because of the singular focus on productivity at the expense of other important systemic dimensions such as the environment or health. The complexity of current challenges requires us to understand them from different perspectives and to revisit the way we think, learning to consider broader social, economic and environmental objectives, which do not necessarily converge or are entirely compatible. Such broader goals require new constellations of alliances, with the actors in charge of defining the goals and specifications of the innovation sought. From this perspective, it seems important to develop a vision in terms of the 'ecology' of innovation, whereby the innovation process would be addressed analytically, focusing on the interactions among its different elements, on the relationships with its environment and on transformations generated by the dynamic complex of interactions and interdependencies.

But a number of questions remain: can we combine different innovation models to bring about synergistic diversity or are some models incompatible? Can we build on what has already been developed, so as not to lose advances to date and extant points of reference? Should we turn towards incremental novelties (gradual change) or radical ones (change everything at once)? While it is clear that actors in each field will have to find their own answers, it is important to recognize that these questions invoke the matter of balance between creative novelty and a destabilizing creativity.

After all, innovation also constitutes a challenge to what already exists, as Juliana Santilli shows (Chapter 9). And in the face of challenge, there is inevitably resistance; every innovation will encounter resistance, which otherwise may have remained latent or hidden. How can we anticipate 'false leads' and avoid the associated risks, disorder and damage? Can we select – in a timely manner – those paths of innovation that will synergize and mobilize actors and inspire them to rise above their differences? Innovation in this way can be seen as a power struggle between conflicting visions and requires us to reflect upon what underlies these visions, or even to create a new vision that will facilitate the search for common goals. Discussing these visions is the first step towards identifying the criteria for choice of processes of change that we want to foster and engage in.

While assessments of existing models of innovation are relatively advanced, covering many sectors and activities from diverse regions, and highlighting processes of exclusion, environmental problems or economic impasses, analyses of new proposals must be encouraged, in order to help other researchers and society more broadly to take decisions. We must give ourselves the means of challenging conventional wisdom in order to design well-reasoned alternatives along with an improved understanding of their promise and limitations. We must avoid politically correct or 'right-minded' thinking: do cooperatives really help small farmers become more secure? Do biotechnologies really lead to the concentration of wealth and increase disparities? Many local experiences lead to new proposals and deserve to be analyzed in greater depth in order to draw general lessons about agricultural innovation.

But it is also a matter of undertaking exploratory approaches, and not only of using what we already know to meet the concerns of the moment. We must learn to explore the unknown, to imagine, to create! Prospective scenario methods, for example, can be used to explore possible changes. By characterizing future states, desired or undesired, foresight can do even more; it can help identify, in advance, the variables that determine the choice of a particular path to take if we prefer one 'future' over another.

12.3 Engaging transitions to generate change

The advent of participatory approaches has accompanied and, in part, driven considerable methodological diversification in innovation research over the past 15 years, whereby ready-made recipes have been abandoned, and new methods advocated supporting collective decision making. Today many tools and methods are available to conduct joint analyses that go beyond simple causes and effects (Why is it unsatisfactory? What makes it critical?), and explore possible futures and desirable outcomes by all the actors (even if the results of these consultative processes are far from guaranteed given the often strongly divergent interests of the actors). And yet, the main challenge today is to help define how to foster the transition from the current reality to the imagined one.

Transition have long been a subject of interest among historians, agronomists, sociologists, but their inner process dynamics often remain obscured between contextual determinism and 'determinant' leaders. In any transition various trajectories are possible in moving from one state to another, and such trajectories are themselves determined by the transition underway. How then to characterise transitions? How to generate desirable dynamics? What changes to encourage? How to adapt to the inherent uncertainties? We need to embark in a process of 'innovative design' and not just of 'controlled-design'.[2]

Innovative entrepreneurs are important in transitions and deserve to be encouraged and allowed to develop their ideas and creativity, as pointed out Andy Hall and Kumuda Dorai (Chapter 4). Niches where innovation can emerge under safe conditions, beyond the constraints of the dominant prevailing model, can allow the entrepreneurs' creativity to blossom. But when the time comes for these innovations to generate a transition to another

[2] We use here the terms devised for the industrial innovation processes by: Le Masson P., Weil B. et Hatchuel A. (2008) Les processus d'innovation, conception innovante et croissance des entreprises. Hermès Lavoisier (Paris), 471 p.

regime[3], an array of challenges arise, concerning scalability, institutions, individual and collective capacities, physical and cultural barriers, etc.

Thus, in agriculture, a key question is that of the social capacity for new technological choices to emerge (given accompanying social, economic and geographical changes), as pointed out by Jean-Marc Meynard (Chapter 5). Past choices may prove difficult to reverse if they have been integrated not only into currently used technical solutions (mechanization, fertilizers, pesticides, genetics, etc.) but also into cognitive systems (knowledge and knowhow, representations of nature, of pollution, of landscapes, etc.) and the value systems of the main actors involved. Do we not risk being trapped by technical rationalization, a sort of 'lock-in', as experienced by other actors elsewhere, in other economic sectors? How can we improve our analysis of these 'lock-ins' so that we can get past them?

Methodological work is indeed here; and the challenge is to identify the concepts, knowledge and skills required for us to envisage the transition. Not to actually determine the transition in advance but to be capable of choosing it as we make our way, to adapt ourselves to arrive eventually at our goal – or, more likely, to continually define new and diverse goals. We have to strive to reach beyond safe targets which do not entail major transformations of current practice and values. We must realize that there are no optimal solutions and the best we can do is to equip ourselves with the means to assess different options as and when they appear, and to offer the available choices to the various stakeholders recognizing the inherent tradeoffs. It is this knowledge of criteria and thresholds that is critical; this skill of assessing and questioning, that we must help foster. It is a learning process that every society engages in, because although choices will have to be made by policymakers, informed by research, they must ultimately depend on society's expectations. It is by creating the conditions for this social learning[4], of knowledge and values consciously built and chosen by society, that we will create the propitious conditions for transitions to other agricultural development models.

Social learning is not something that necessarily happens naturally or smoothly, nor does it arise out of consensus: it requires social debate, a clash of opinions and visions. How will the concepts and knowledge that may guide the transition be produced and legitimized? Through which networks, within which dominant or marginalized institutions, and through which processes will this ocurr? Concepts that guide thinking, objects that focus attention, indicators and criteria that enable evaluation, chosen thresholds and envisioned processual

[3] In transition management theory, "regime" designs a stable state in which the dominant practices, rules and technologies reinforce a particular socio-technical system. "Niche"is an 'area' in which particular conditions provide space for radical innovation and experimentation. For further development, refer to: Kemp, R; Loorbach, D and Rotmans, J (2007). "Transition management as a model for managing processes of co-evolution towards sustainable development". The International Journal of Sustainable Development and World Ecology 14 (1): 78–91.

[4] The premises of social learning applied to agriculture were first discussed by Rôling & Wagemakers as a way of dealing with the unpredictability and complexity of making the shift towards an ecological knowledge system: Röling, N., Wagemakers M., 1998. Facilitating Sustainable Agriculture: Participatory Learning and adaptive management in times of environmental uncertainty. Cambridge University Press, Cambridge, New York, Melbourne.

steps to be taken are all part of a cognitive exercise which invites us to reflect critically and recursively, and which will gradually drive social learning.

12.4 Innovating is primarily learning through interactions

How can social learning in innovation be rendered? What type of interactions and structures will lead to innovation? Working together in spite of our differences already represents a major challenge. And innovation, as we have said before, does not concern only researchers; farmers, rural stakeholders, policymakers, scientists – and all the diversity behind each of these 'categories' – have to discuss and interact. Fortunately, institutions do exist to facilitate and formalize these exchanges, but often interactions get bogged down in routines and the actors lose sight of the main objective. We must give ourselves the means to reinvigorate dialogue and ensure flexibility so that a collective creativity can develop and we are able to imagine new solutions. To do so, unifying concepts are necessary to get people to discuss together within innovation systems, as argued by Bernard Triomphe and Rikka Rajalahti in Chapter 3. Such concepts usually develop within sectors or regions by focusing on a specific 'problem'. However, there is also much to be gained from discussions across different types of sectors, at different scales.

These discussions, we observe, often take place outside the ambit of well recognized institutions, on the margins of formalized systems, where there is room for more creativity and spontaneity. Surprises should emerge; risk-taking must be possible. That is why formal or informal networks that serve as platforms for exchange are so important. Intermediaries, mediators and facilitators that bridge institutions and social groups play a fundamental role – assuming, of course, that they do encumber the very processes they are supposed to facilitate. They translate ideas from one group to another, find common interests, 'hybridize' knowledge and mobilize people from different groups. They create links between different worlds, open up new areas of sharing, and exchange in spaces where new ideas and potential innovations can foment and flourish.

This does not mean that more formal institutions are not important in their own right. The State, private firms, civil associations, among others, can create propitious conditions for innovations to emerge and also open up new spaces, as Karim Hussein and Khalid El-Harizi point out (Chapter 10). And, once innovations do appear, these institutions can formalize the links among actors through contracts, regulations, agreements and resources to help innovations spread more widely. Supply chains and private actors have a particularly important role to play in this type of 'up-scaling' by providing some measure of security to offset risk taking which often accompanies innovation adoption. But it is important for them to be built collectively so that effects of exclusion can be avoided, as highlighted by Estelle Biénabe, Cerkia Bramley and Johann Kirsten (Chapter 7).

Clearly, creators and formalizers[5] in the innovation process have different roles according to their respective capacities. How can we encourage the comparative advantage of a diversity of actors and help them build specific skills? How should resources (financial, social, and institutional) be directed to different types of actors? These questions arise because innovation often brings with it conflicts over distribution or social exclusion. Clearly, some actors remain on the sidelines in innovation processes; that is not the problem. The issue is whether or not actors have the choice to participate in such processes. How can we reduce the asymmetries that arise in the process so as to facilitate exchanges among equals? How can propitious conditions be created to encourage participation by women, young people or immigrants in innovation? As illustrated by Denis Requier-Desjardins (Chapter 6), the debate on policies for innovation cannot be considered in isolation from other policies, for example, those guaranteeing fundamental rights, education, health care and access to infrastructure. The 'recruiters' in innovation processes, be they leaders of an association or region, entrepreneurs, or researchers, can also play an instrumental role. They are the ones who can ensure the participation of marginalized actors, convince them of their importance within the group, strengthen their capacities to contribute and adapt – even to co-govern – the innovation process, thus leading to the democratization of innovation processes. Power relations do not, of course, disappear completely but they can be partially counterbalanced or, at least, revealed. André Torre and Frédéric Wallet (Chapter 8) show the importance of conflicts within innovation processes.

12.5 The researcher, an agent of change?

Following our consideration above of the nature of agriculture, visions for the future, transitions and the social interactions necessary to implement them, it would seem important to reflect upon our own role: how can researchers participate in the development and diffusion of innovation? Unfortunately, creativity is not always a strong point of scholarly research. And yet, research can play a significant role if researchers are able to self-reflection and self-analyze. The role of research is less to provide ready-made answers as it is to help define the problem, support decision-making in the innovation process and help co-construct new knowledge.

The (re)definition of the problem can be undertaken through a research that involves denaturalization/regeneration, by the deconstruction of largely accepted evidence, by the uncovering of known but unrecognized facts and unperceived facts, and through the creation of knowledge and new relationships. This endeavour ought to bring actors to question the standards, concepts and efficiency criteria used so that they can be transformed according to the changes underway.

[5] Although many studies have used these concepts, the forefather is of course : Rogers E.M., 1962. Diffusion of Innovations, New York, Free Press.

The research process is not separable from decision-making: investigation is central to the design of solutions, requiring explicit choices to be made. By assuming our role of clarifying the implications of different choices, researchers must recognize the political role of social research, potentially empowering actors in innovation. Indeed, faced with prevailing development models, which are taken as being self-evidently valid, it is a challenge for citizens to design alternatives that are not based almost exclusively on these ideological foundations. Science can be expected to provide a range of models in support of social choices to be made in a democratic society. It must be remembered, however, that answers provided by Science are shaped by theoretical frameworks and paradigms that are not necessarily definitive.

Research can also help balance the relationships between different forms of knowledge, by placing them in their context, considering their scope and assessing their legitimacy. This work is a prerequisite to discussions between different viewpoints so as to avoid privileging only one side. In this way, citizens can be helped to understand their own representations and those of others, to overcome them and to build together new concepts and knowledge to deal with their problems. By engaging with others in collective action, research becomes involved in the simultaneous reconstruction of the knowledge of actors and the knowledge of research, either by revealing existing knowledge or producing new knowledge. This reconstruction occurs on the one hand for the concepts which enable to think together about the goals aimed at, the objects to reach these goals and the tasks to implement it, and on the other hand, the knowledge which will enables to achieve these goals.

Through this tripartite approach – of problem definition, clarification and interaction – researchers can help build reflexivity in the innovation research process that will allow for critical reflection on the objectives and means used, i.e. an ongoing evaluation, which is a condition for social learning. For this to occur, we must devise performance criteria that can be used to evaluate these new models in terms of the multiple parameters to be considered (productivity, employment, ecosystem services, climate change, development of rural areas, etc.). Such criteria will be indispensable for the development of new agricultural production models and to manage them, and to ensure that different types of agriculture complement one another.

Change entails a social dynamic in which not only the perspectives and objects of interest are transformed, but also the actors themselves in the collective learning process. As Papa Seck, Aliou Diagne and Ibrahima Bamba show in Chapter 11, the success – and thus the value – of the involvement of researchers rests on their ability to engage in such learning. Not all have this ability; it depends on their scientific culture, their position in the group and their capacity to engage in a process whose end goal is not known at the outset. In any exploratory research, the elements identified as being important for decision making are the result of the process and not its starting point. Such an approach requires a full consideration of other possible worlds through joint discussion of construed facts and values. It is by considering together distinct rationalities, knowledge, facts, codes, standards, values and hybrid actor networks that innovation can develop. And researchers do not usually work in this way, i.e. they follow particular paradigms in which facts are separated from values. How many among us are actually prepared to engage in social learning processes from which we will emerge inevitably transformed, given the inseparability of knowledge and relationships?

How can we be supported in this risky journey, to identify procedures that ensure that our contributions are rigorous and our alliances legitimate? Perhaps the answer lies in the experience of this symposium, i.e. by drawing on a diversity of experiences from others, by learning from our mistakes and not being satisfied by success stories but by analyzing in greater depth the failures. And, of course, we must continue working on developing conceptual frameworks, approaches and methods that ensure effective involvement in the dynamic process of change.

These areas of research need urgent attention and investment by organizations engaged in research on agricultural innovation and development, so that they are not left solely to the initiative of individual researchers. Here too, the paths to success are many and must be adjusted to a variety of situations, problems and scientific cultures. Our hope is that symposia such as ISDA 2010 confirm the broad interest in developing effective innovation systems and highlight the ways in which researcher involvement in the collective endeavour can drive change! Let us indeed be *en route*, but not before making an appointment for an update a few years hence!

About the editors

Emilie Coudel
Agro-economist at Cirad, Dr. Emilie Coudel is interested in the capacities required for transitions, involving different types of innovation processes. Her studies focus on empowerment and capacity building of farmers to encourage their participation in governance, with emphasis on environmental management in Brazil. She was project manager for the ISDA symposium.
UR Green
Cirad
Campus International de Baillarguet
34398 Montpellier Cedex 5
France
E-mail: emilie.coudel@cirad.fr

Hubert Devautour
Dr. Hubert Devautour is an agricultural- and food-industries engineer and doctor in agrifood economics. He has worked on the valorisation of production of family farms and has studied innovation processes in artisanal food production in Africa and in improving quality of food products. He is deputy director of the 'Environment and Societies' science department of the Centre for International Cooperation in Agronomic Research for Development (CIRAD). He was chair of the organizing committee of the ISDA symposium.
Cirad
Campus International de Baillarguet
34398 Montpellier Cedex 5
France
E-mail: hubert.devautour@cirad.fr

Guy Faure
Dr. Guy Faure is an economist at CIRAD, deputy director of the UMR Innovation joint research unit. He has undertaken research on farming systems, advisory services and farmers' organizations using action-research approaches in Africa and Latin America.
CIRAD
UMR Innovation
73 avenue Jean-François Breton
349398 Montpellier Cedex 5
France
E-mail: guy.faure@cirad.fr

Bernard Hubert

Prof. Dr. Bernard Hubert is director of research at INRA and Director of Studies at EHESS. Formally trained as an ecologist, he is interested in scientifically assessing and taking into account human activities. He is also interested in the social sciences and their epistemological contribution to the life sciences. This led him to focus on interdisciplinarity and how to conduct research to solve problems in close interaction with the concerned stakeholders. The concept of sustainable development has, in recent years, provided the intellectual framework for the notion of integration (of disciplines, social actors, public and private actions) by proposing a deliberate manner of designing and conducting research in society. Since April 2009, he heads Agropolis International in Montpellier. He was chair of the scientific committee of the ISDA symposium.

Agropolis International
1000, avenue Agropolis
34394 Montpellier Cedex 5
France
E-mail: bernard.hubert@agropolis.fr

Christophe-Toussaint Soulard

Dr. Christophe-Toussaint Soulard is a geographer and associate researcher at INRA. His research focuses on innovative agricultural systems in rural and periurban areas. He is in charge of multidisciplinary participatory research projects, in France, the Mediterranean region and western Africa. He is the head of the UMR Innovation joint research unit (http://umr-innovation.cirad.fr/) in Montpellier, France.

INRA UMR Innovation
2 place Viala
34060 Montpellier cedex 1
France
E-mail: soulard@supagro.inra.fr

About the authors

Ibrahima Bamba
Dr. Bamba is currently working for IFAD. Prior to that he was with Africa Rice from 2007 to August 2011. He obtained a PhD in Agricultural Economics from the University of Kentucky in 2004 and a 'Maitrise' in Economics from the University of Aix-Marseille II.
International Fund for Agricultural Development
Via Paolo di Dono 44
00142 Rome
Italy
E-mail: i.bamba@ifad.org

Estelle Bienabe
Dr. Estelle Biénabe is an Agricultural Economist at CIRAD. She has expertise in market access and quality dynamics and in environmental policy analysis.
CIRAD
TA 179 / 04
Avenue Agropolis
34398 Montpellier cedex 5
France
E-mail: estelle.bienabe@cirad.fr

Cerkia Bramley
Cerkia Bramley is a researcher at the International Development Law Unit of the University of Pretoria. A qualified attorney with a background in agricultural economics, her research interests include issues related to intellectual property and food law.
Room 2-30
Department Agricultural Economics and Rural Development
Agricultural Annex
University of Pretoria
South Africa
E-mail: cerkia.bramley@up.ac.za

Lawrence Busch
Prof. Dr. Lawrence Busch is University Distinguished Professor of Sociology at Michigan State University. He has published more than 150 publications including *Standards: Recipes for Reality* (MIT Press, 2011). He is a member of the French Académie d'Agriculture.
Department of Sociology
Michigan State University
East Lansing, MI 48824
USA
E-mail: lbusch@msu.edu

Patrick Caron

Patrick Caron is currently Deputy Director General for research and strategy at CIRAD. Veterinarian, with a PhD in geography, he specialized in family farming, farming and livestock systems and territorial development, themes on which he has coordinated many scientific projects. His career abroad was mainly conducted in Brazil and southern Africa. At CIRAD since 1988, he was Deputy Director for Scientific Affairs at the Department Tera (Territories, Environment and People) from 2001 to 2004 and Director of the Department Environments and Societies (Es) from 2007 to 2010.

CIRAD
Campus international de Baillarguet
TA C-DIR/B, 34398 Montpellier Cedex 5
France
E-mail: patrick.caron@cirad.fr

Olivier T. Coomes

Oliver T. Coomes is a Professor of Geography at McGill University. His research examines peasant livelihoods, poverty and environmental change among traditional peoples of Amazonia and other regions of the Neotropics.

Department of Geography
Burnside Hall, Rm 705
McGill University
805 Sherbrooke Street West
Montreal, Quebec H3A 0B9
Canada
E-mail: oliver.coomes@mcgill.ca

Aliou Diagne

Dr. Aliou Diagne is a specialist in Agricultural Economics and in Economics. He has been serving as associate editor for the *African Journal of Agricultural and Resource Economics* (2005 to present) and for *Food Security* (2009 to present). He is currently the programme Leader for Policy, Impact Assessment and Innovation Systems.

01 BP 2031
Cotonou
Benin
E-mail: a.diagne@cgiar.org

Kumuda Dorai

Kumuda Dorai is Programme Officer at LINK Ltd, a company providing research and advisory services on innovation in development, which started off as a UNU-MERIT/FAO initiative. Her professional interests include exploring ways to go beyond traditional notions of development assistance. Her work at LINK also involves research on new socially relevant sources of innovation, including emerging trends in hybrid entrepreneurship.

LINK Ltd.
197 Osborne Road
Brighton, BN 1 6LT
United Kingdom
E-mail: kumuda.dorai@innovationstudies.org

Khalid El Harizi

Khalid El Harizi, Agricultural Economist, joined the International Fund for Agricultural Development in 1991. Currently managing the Cambodia Programme, he has a keen interest and solid experience in the role innovation can play in empowering the rural poor.

IFAD
Via Paolo di Dono, 44
00142 Rome
Italy
E-mail: k.elharizi@ifad.org

Andy Hall

Professor Andy Hall is Director of LINK Ltd, a company providing research and advisory services on innovation in development. He was previously senior research fellow at UNU-MERIT and he is currently visiting professor at the Open University, UK. His professional interests include the application of systems ideas in agricultural development and capacity building. His current research is exploring new socially relevant sources of innovation in emerging economies.

LINK Ltd.
197 Osborne Road
Brighton, BN 1 6LT
United Kingdom
E-mail: andy.hall@innovationstudies.org

Jean-Marc Meynard

Dr. Jean-Marc Meynard is director of research at the French National Institute for Agricultural Research (INRA) and has been head of its 'Sciences for Action and Development' research department from 2003 to 2012. His research deals with the design of innovative agricultural systems, with the objective of reducing inputs and conciliating productive and environmental performances. Since 2008, he has been chair of the French Scientific Council for Organic Agriculture.

INRA-SAD
Route de Saint-Cyr
78026 Versailles Cedex
France
E-mail: meynard@grignon.inra.fr

Karim Hussein
Karim Hussein is Technical Advisor for the Africa Partnership Forum Support Unit, OECD. He has worked for more than 20 years on development policy and practice, focusing on livelihoods, poverty, access to innovation, demand-led agricultural services, food security, participation, producer organizations and conflict. He coordinated programmes on fostering partnerships for agricultural innovation in Africa at the OECD and IFAD.
OECD
2, rue André-Pascal
75775 Paris
France
E-mail: karim.hussein65@yahoo.co.uk

Johann Kirsten
Dr. Johann Kirsten is Professor and head of the Department of Agricultural Economics, Extension and Rural Development at the University of Pretoria. He served as the Vice-President of the International Association of Agricultural Economists for the period 2006-2009.
Department of Agricultural Economics
Extension and Rural Development
University of Pretoria
South Africa
E-mail: johann.kirsten@up.ac.za

Riikka Rajalahti
Dr. Riikka Rajalahti works as a Senior Agricultural Specialist in the Agriculture and Rural Development department of the World Bank, specializing in Agricultural Innovation Systems (AIS). Her main task in this role is to promote operationalization of the AIS approach in the Bank's lending operations through the capture of global good practices, knowledge transfer and direct cross-support to regions. One of her recent key activities has been to lead the development of an Investment Sourcebook on AIS.
The World Bank
Department of Agriculture and Rural Development
MS5-514
1818 H Street, NW
Washington DC 20433
USA
E-mail: rrajalahti@worldbank.org

Denis Requier-Desjardins
Prof. Dr. Denis Requier-Desjardins is interested in development economics, rural development (particularly in Latin America), localized agrifood systems, migration, remittances and development. Formerly at the universities of Constantine (Algeria), Abidjan (Côte d'Ivoire), Lyon Rennes and Versailles, he is currently at the IEP of Toulouse, University of Toulouse.
Institut d'Etudes Politiques de Toulouse (Sciences-po Toulouse)
2ter rue des Puits Creusés
31685 Toulouse Cedex 6
France
E-mail: denis.requier-desjardins@univ-tlse1.fr

Juliana Santilli
Dr. Juliana Santilli is a lawyer, public prosecutor and researcher, specialized in environmental and cultural heritage law and public policies. Author of the book *Agrobiodiversity and the Law: regulating genetic resources, food security and cultural diversity* (Earthscan, 2012).
S.Q.N. 213
Bloco K, apto. 101
Asa Norte-Brasília, DF – cep 70.872-110
Brasil
E-mail: juliana.santilli@superig.com.br

Papa Abdoulaye Seck
Dr. Papa Abdoulaye Seck is a specialist in agricultural policy analysis and strategy. He is a member of the Senegal Academy of Sciences and has the rank of Director of Research. Since 2006, he is Director General of AfricaRice, a CGIAR Consortium Research Center.
Director General
Africa Rice Center (AfricaRice)
01 B.P. 2031 Cotonou
Benin
E-mail: p.seck@cgiar.org

André Torre
Prof. Dr. André Torre is research director at INRA (National Institute of Agronomical Researches) and AgroParisTech in Paris. He teaches in several French Universities. He is chief editor of the *Revue d'Economie Régionale et Urbaine* and immediate past-President of the French speaking section of ERSA. He has published several books and papers on the topics of proximity, innovation and regional development.
UMR SAD-APT
Agro Paristech
16, rue Claude Bernard
75231 Paris Cedex 05
France
E-mail: torre@agroparistech.fr

Bernard Triomphe
Dr. Bernard Triomphe is a systems agronomist with 25 years of experience in developing countries. His interests include assessing farming systems and innovation processes, and contributing to designing innovations as part of multi-stakeholder partnerships.
CIRAD UMR Innovation
TA C-85/15
73, rue Jean-François Breton
4398 Montpellier Cedex 5
France
E-mail: bernard.triomphe@cirad.fr

Frédéric Wallet
Dr. Frederic Wallet is an economist whose work is devoted to the analysis of institutional innovation in rural development policies and territorial governance. His secondary research theme focuses on the evolution of regulatory mechanisms for GI products and their impact on territories.
INRA
UMR SADAPT
16 rue Claude Bernard
75321 Paris Cedex 5
France
E-mail: wallet@agroparistech.fr

Authors of boxes

Rodolfo Araya, Universidad Costa Rica, San José, Costa Rica
Didier Bazile, Cirad, UPR 47/Green, Instituto de Geografi a, PUC Valparaíso, Chile
Mélanie Blanchard, Cirdes, 01 Bobo-Dioulasso, Burkina Faso
Ochieng Bolo Maurice, The Open University, Walton Hall, Milton Keynes, MK7 6AA, United Kingdom
Philippe Bonnal, Cirad, UMR Politiques et Marchés, Montpellier, France
François Boucher, Cirad, UMR Innovation, IICA-México, Mexico
Virginie Brun, IICA-México, Mexico
Shambu Prasad Chebrolu, Xavier Institute of Management, Bhubaneswar, Orissa, India
Eduardo Chia, Inra Supagro Cirad, UMR Innovation, Montpellier, France
Hubert De Bon, Cirad, UPR Hortsys, Montpellier, France
Noya Eliane de Carvalho, Instituto Agronomico de Pernambuco IPA, Pernambuco, Brasil
Ludivine Eloy, UMR Art-Dev 5281, Montpellier, France
Laure Emperaire, UMR 208 Patrimoines locaux, France
Kebebe Ergano, International Livestock Research Institute, Addis Ababa, Ethiopia
Mostafa Errahj, École nationale d'agriculture de Meknès (ENA), département Ingénierie du développement, Meknès, Morroco
Edith Fernandez-Baca, The Consortium for the Sustainable Development of theAndes Eco-Region, Calle Mayorazgo 217, San Borja, Lima 41, Peru
Graciela Ghezán, EEA INTA, Balcarce, Argentina
Henri Hocdé, Cirad, Avenue Agropolis, Montpellier, France
Zakaria Kadiri, Laboratoire Méditerranéen de Sociologie (Lames), UMR G-eau, Aix en Provence, France
Rémi Kahane, GlobalHort, c/o AVRDC, RCA, PO Box 10 Duluti, Arusha, Tanzania
Mahamoudou Koutou, Cirdes, 01 BP 454 Bobo-Dioulasso 01, Burkina Faso
Marcel Kuper, Cirad, UMR G-Eau, Avenue Agropolis, Montpellier, France
Pieter Lemmens, Centre for Methodical Ethics and Technology Assessment (META), Wageningen University, The Netherlands
Allison Loconto, Department of Sociology, Michigan State University, 422b Berkey Hall, USA
Bezerra Lopes Geraldo Majella, Instituto Agronomico de Pernambuco IPA, Pernambuco, Brasil
Mónica Mateos, Fac. Ciencias Agrarias, UNMdP Ruta 226, km 73,5 Balcarce, Argentina
Laurent Parrot, Cirad, UPR Hortsys, TA B-27/PS4, Boulevard de la Lironde, Montpellier, France
María Paz-Montoya, The Consortium for the Sustainable Development of the Andes Eco-Region, Calle Mayorazgo 217, San Borja, Lima 41, Peru
Philippe Pédelahore, Cirad, UMR Innovation, RAD, Centre de Nkolbisson, BP 2076 Yaoundé, Cameroon
Marc Piraux, Cirad, UMR Tetis, Universidade Federal de Campina Grande, Brasil

Khaldi Raoudha, Institut national de la recherche agronomique de Tunisie, rue Hédi Karray 2080, Ariana, Tunisia

Hélène Rey-Valette, Université Montpellier 1, UMR Lameta, Montpellier, France

Juan Carlos Rosas, Escuela Agrícola Panamericana, Zamorano, P.O. Box 93, Tegucigalpa, Honduras

Bernard Roux, CESAER/Agrosup Dijon, France

Guido Ruivenkamp, Wageningen University, Department of social sciences, Critical technology Construction, Hollandseweg 1, 6706 KN Wageningen, The Netherlands

Emmanuel Simbua, Tea Research Institute of Tanzania, P.O. Box 2177, Dar es Salaam, Tanzania

Max Thomet, CET-Sur, Temuco, Chile

Eric Vall, Cirad, Avenue Agropolis, Montpellier, France

Natalia Yañez, The Consortium for the Sustainable Development of the Andes Eco-Region, Calle Mayorazgo 217, San Borja, Lima 41, Peru

Printed in the United States
by Baker & Taylor Publisher Services